AF601871

Basics of
Plant Breeding

The Author

Dr. Hasan Khan did his Ph. D. in the discipline of Genetics and Plant Breeding during the year 2011 from GKVK, UAS, Bengaluru. He is presently working as Scientist (Plant Breeding), AICRP on Groundnut, MARS, UAS, Raichur.

Dr. Hasan Khan has significantly involving and contributing in the field of crop improvement. During his M.Sc. graduation he worked on Rice (Chemical Induction of Male Sterility), while his Ph. D. research on cowpea crop (Assessment of Morphological and Molecular Diversity and Genetics of Resistance to Bacterial leaf blight and Mosaic virus in Cowpea). When he joined at UAS, Raichur during 2011, he started working on Cotton improvement & collaborated for a varietal release, along with that he also involved in development of a Pigeon pea variety (in 2015). He started his work on Groundnut improvement during 2013 & to cater the immediate needs of farmers adopted two high yielding groundnut variety during 2015. Presently, he is focusing on development of high yielding foliar disease resistance groundnut varieties suitable Hyderabad-Karnataka region.

Dr. Hasan Khan received several fellowships during his school days as well as in higher studies and awards from the professional societies for contributions in the field of crop improvement. He has published more than 20 research articles in reputed National & International Journals. He also handled five projects (2 completed, 3 ongoing) with total outlay of more than 100.0 lakhs and financed by GoK, RKVY & BMFG through ICRISAT. Presently he has six M.Sc. students working under his chairmanship and serving as member of advisory committee for more than 20 P.G. students of UAS, Raichur.

Basics of
Plant Breeding

— *Author* —

Dr. Hasan Khan

Scientist (Plant Breeding),
AICRP on Groundnut,
MARS, UAS,
Raichur, Karnataka

2020

Daya Publishing House®

A Division of

Astral International Pvt. Ltd.

New Delhi – 110 002

ISBN 9789387057234 (Int. Edition)

Publisher's Note:

Every possible effort has been made to ensure that the information contained in this book is accurate at the time of going to press, and the publisher and author cannot accept responsibility for any errors or omissions, however caused. No responsibility for loss or damage occasioned to any person acting, or refraining from action, as a result of the material in this publication can be accepted by the editor, the publisher or the author. The Publisher is not associated with any product or vendor mentioned in the book. The contents of this work are intended to further general scientific research, understanding and discussion only. Readers should consult with a specialist where appropriate.

Every effort has been made to trace the owners of copyright material used in this book, if any. The author and the publisher will be grateful for any omission brought to their notice for acknowledgement in the future editions of the book.

Published by : **Daya Publishing House®**
A Division of
Astral International Pvt. Ltd.
– ISO 9001:2015 Certified Company –
4736/23, Ansari Road, Darya Ganj
New Delhi-110 002
Ph. 011-43549197, 23278134
E-mail: info@astralint.com
Website: www.astralint.com

Digitally Printed at : **Replika Press Pvt. Ltd.**

Foreword

The science of plant breeding has advanced considerably making rapid strides in 20th century. The accumulation of knowledge on the basis of research conducted all over the world particularly in the related discipline like Genetics, Bio-chemistry and Molecular Biology have led to refinement of crop improvement approaches. The notion that the plant Breeding is "an art and science" which was there earlier has indeed now become a science.

The refinements of these approaches have led to the improvement in accuracy and efficiency of crop breeding techniques. Besides, while treating the crop improvement approaches in details in any such text books. It is all the more necessary to give due importance in describing all the related aspects to make any such volume complete and meaningful. For instance, the scientific basis of any approach, like principles of Biometric genetics. Genetics, Seed Science, Molecular Biology and exposure to the more important topics of current interest like IPR, etc., is very critical. All these have been taken good care of by the author of this book. Although the science of the plant breeding is no more as simple as it was, it is absolutely necessary to understand the basis principle of Plant Breeding and closely related sciences like genetics. Only a good comprehension of these principles would enable anybody to understand all the complex issues which are now the basis of the present approaches. The author has done a good job in bringing these aspects also very clearly. The contents of the book are therefore very comprehensive and informative which will be quite useful to the students.

The author besides being highly qualified in the subject of Plant Breeding has vast experience in practical of Plant Breeding and has been teaching science of Plant Breeding to the UG and PG students in this University. Therefore, his unique experience of theoretical knowledge and practical Plant Breeding has culminated in Publication of this useful book.

(P.M. Salimath)
Vice-Chancellor, UAS, Raichur

Preface

Crop improvement is essential and continuous tasks for increasing agricultural productivity. No amount of production practices can make crop varieties perform beyond its genetic potential. The knowledge of plant breeding is not only helps in increasing crop yields and quality but also providing means for incorporating resistance into crop plants against biotic and abiotic stresses. Plant breeding is offered as a subject at under and post graduate studies in all SAU's in India. The objective of writing the present book ***'BASICS OF PLANT BREEDING'*** is to help the Indian students, especially the under graduate to understand the subject in the simplest way.

This book covers the syllabi offered by various Indian universities (SAU's and Conventional) at UG and PG levels. It also covers the syllabi of JRF, SRF, ARS and NET examinations conducted by ICAR, New Delhi, in the concise manner. It includes several additional important topics like extensive history of plant breeding and developments; modes of reproduction and pollination; self-incompatibility and male sterility; domestication and PGR management; breeding methods for self, cross and asexually propagated crops; biometrics in plant breeding; wide hybridization; mutagenesis in plant breeding; heterosis and inbreeding depression; ideotype breeding; polyploidy and crop improvement; breeding for resistance to abiotic and biotic stresses; plant breeders, farmers rights and privileges; PPV & FR act 2001; participatory plant breeding; modern plant breeding strategies and other important breeding concepts.

Hope this book will be useful for the students, lecturers and researchers in the field of crop improvement. It will also be useful for those appearing in competitive examinations such as JRF, SRF, ARS and NET. Suggestions for further improvement of this book shall be highly appreciated.

I express sincere thanks to Daya Publishing House a division of Astral International Pvt. Ltd., New Delhi for publishing this manuscript.

Dr. Hasan Khan

Contents

List of Figures

List of Tables

1

Introduction to Plant Breeding

Definitions

Plant breeding can be defined as an art, a science, and technology of improving the genetic makeup of plants in relation to their economic use for the mankind.

OR

Plant breeding is the art and science of improving the heredity of plants for the benefit of mankind.

The art of plant breeding lies in the breeder's skill in observing plants with unique economic, environmental, nutritional, or aesthetic characteristics. Before plant breeders possessed the scientific knowledge that is available to them today, they relied solely on skill and judgement in selecting novel plants which could be propagated through seeds or vegetative parts. Plant breeding developed into a science as knowledge progressed in classical genetics and related plant sciences. The foundation of plant breeding was based on recognition of the gene as the unit of heredity, on procedures for gene manipulation, and on rules of genetic behaviour that permitted accurate prediction of the results from gene manipulations.

History and Development of Plant Breeding

- ☆ The modern plant breeding methods have their basis in the genetic and cytogenetic principles. Numerous workers who determined the various modes of inheritance have contributed to the development and understanding of plant breeding.

Table 1: Selected Milestones and Development of Plant Breeding

Year	*Scientist*	*Contribution*
9000 BC	–	First evidence of plant domestication in the hills above the Tigris river (between Iraq and Iran).
3000 BC	–	Domestication of all important food crops in the Old World completed (Wheat, Barley, Peach).
1000 BC	–	Domestication of all important food crops in the New World completed.
700 BC	Babylonians and Assyrians	Pollinated date palm artificially.
1694 AD	R.J. Camerarius (aka Rudolph Camerer, Germany)	First fellow to report sexual reproduction in plants (Maize). He established that seed cannot be produced without participation of pollen.
1716	Mather (USA)	For the first time observed natural crossing in maize.
1717	Thomas Fairchild (an Englishman)	Conducted an interspecific cross (a cross between two species) between sweet william (*Dianthus berbatus*) and *D. Caryophyllis*, to obtain what became known as Fairchild's sweet William/Fairchilds mule.
1727	Vilmorins	✰ The first plant breeding company was established in France ✰ Vilmorin Company of France introduced the pedigree method of breeding
1752	Haartman	Described a natural hybrid in *Trillium* (an ephemeral plant).
1753	Carolus Linnaeus (1707–1778, Swedish botanist)	The landmark work which culminated in '*The binomial systems of classification*' of plants is invaluable to modern plant breeding.
1760-1766	Joseph Kolreuter (1733-1806)	Made extensive crosses in Tobacco and *Solanum* and studied the progenies in detail.
1705-1816	Sprengel	Provided detailes information about the structure and function of flowers. Identified self-and cross-fertile plants and also the factors like flower color and nector which influenced self-or cross pollination.
1759-1835	Thomas Andrew Knight	The first man to produce several new fruit varieties by using artificial hybridization.
–	David Fife, a farmer, Canada	The famous Red Life variety of wheat was developed from a single rust resistant plant by practicing selection.
1819	Le couteur and Patrik shireff	Used individual plant selections and progeny test to develop some useful cereal varieties
1838	Schleiden and Schwan	Proposed Cell Theory, stating cells to be basic unit of biological organisation.
1843	Le coutier	A farmer published his results on selection in wheat. He concluded that progenies from single plants were more uniform.

Contd...

Table 1–*Contd...*

Year	*Scientist*	*Contribution*
1847	–	"Reid's Yellow Dent" maize (*Zea mays* var. *indentata*) developed. It contains high soft starch, dent: small indentation at the crown of each kernel.
1849	Gartner	Experimented with 700 species belongs to 80 generas and obtained about 250 hybrids out of 10,000 crosses affected by him.
1856	Vilmorin	✰ Proposed individual plant selection based on progeny testing and this was known as '*Vilmorins principle of progeny testing*'. ✰ He proposed this progeny testing in sugar content in sugar beets (*Beta vulgaris*). ✰ But this method was in- effective in wheat. ✰ This clearly demonstrated the difference between effect of selection in cross and self pollinated crops.
1859	Charles Darwin	✰ The 'Theory of evolution through natural selection (Origin of Species by Natural Selection). ✰ The Effect of Cross and Self-fertilization (1889).
1866	Mendel	Published his discoveries in *Experiments in plant hybridization*, cumulating in the formulation of laws of inheritance and discovery of unit factors (genes).
1873	Patrick Shireff	✰ He concluded that only the variation heritable nature responded to selections, and that there variation arose through 'natural sports' (=mutation) and by 'natural hybridization' (=recombination during meiosis in the hybrids so produced).
1882	Fleming	Identification of chromosome and their behaviour in cell division (Mitosis).
1882	Alphonse de Candole	Provoded the first account of history and origin of cultivated plants
1884	Strasurger	Described fertilization in angiosperms. ('1877: Hertwig recognised the role of sperm and egg in fertilization)
1890	Nilson-Ehle *et al.*	At Swedish Seed Association, Svalof Sweden refined the single plant selection.
1890	Rimpu (Sweden)	Affected cross between *Triticum aestivum* x Rye (*Secale cereale*), which later on gave birth to Triticale (First man made cereale).
1899	Hopkins	Described the ear-to-row selection method of breeding in maize.
1900	Hugo de veris, Tshermark and Correns	The science of genetics began with the rediscovery of Gregor Johan Mendel's paper, which was originally published in1866.

Contd...

Table 1–*Contd...*

Year	*Scientist*	*Contribution*
1903	Sutton and Boveri	Demonstrated that genes are situated on chromosomes, as a consequence of which it become possible to locate their exact position for effective manipulation.
1903	Johansen	☆ Proposed the famous 'pure line theory' ☆ It states that a pure line is progeny of a single self fertilised homozygous plant. ☆ He proposed this theory based on his studies in *Phaseolus vulgaris* (common bean).
1904–1905	Nilsson-Ehle	Proposed the multiple factor hypotheses for inheritance of colour in wheat pericarp.
1908–1909	Hardy (England) and Weinberg (Germany)	Developed the law of equilibrium of populations. It is also called as *Hardy-Weinberg* Law.
1908–1910	East	Published his work on inbreeding
1909	Shull	Conducted extensive research to develop inbreeds to produce hybrids. He described in detail about the effect of inbreeding.
1918	Fisher	Successfully applied Mendalian principles to explain the genetic control of continuous variation (Quantitative traits).
1919	D. F. Jones	Took the idea of a single cross further by proposing the double-cross concept, which involved a cross between two single crosses.
1925	N. I. Vavilove	Proposed the concept of **centres of origin**.
1926	–	**Pioneer** Hi-bred Corn Company established as first seed company.
1928	Stadler	Proposition of artificial mutagenesis through X-ray treatment
1928	Griffith	Demonstrated that heriditary material *i.e.* genes are composed of DNA that led to extensive investigation of chemical composition and modes of action of genes.
1930s	–	Era of mutation breeding.
1934	Dustin	Discovered colchicines (chromosome doubling agent).
1935	N. I. Vavilov	Published - *The Scientific Basis of Plant Breeding.*
1937	Blakeslee and Avery	Chromosome doubling in plants by treating with Cochicine
1938	D. Tollenar	First induced mutant variety of Tobacco 'Chlorina' by X-rays irradiation.

Contd...

Table 1–*Contd...*

Year	*Scientist*	*Contribution*
1940	Harlan	Used the bulk breeding selection method in breeding
1944	Avery, MacLeod, and McCarty	Discovered DNA is hereditary material
1945	Hull	Proposed recurrent selection method of breeding
1950	B. McClintock	Discovered the *Ac-Ds* (Activator-Dissociator) system of transposable elements/jumping genes in maize.
1953	Watson, Crick, and Wilkins	Proposed a model (double helix) for DNA structure
1960's	Ake Gustafssen	✰ Developed a Barley mutant varity 'Pallas' by X –ray irradiation. ✰ Father of mutation breeding.
	Norman E. Borlaug	✰ At 1953, for the first time outlined the methods for developing multilines in wheat ✰ Developed Mexican semi dwarf wheat varieties which paved the way for green revolution in wheat. ✰ The dwarfing gene was isolated from wheat variety **Norin 10.**
	Dr. M. S. Swaminathan (1963)	✰ Introduced the Mexican dwarf varieties in the India and a number of high yielding wheat varieties like Kalyan sona, Sharbathi sonara were developed.
	–	✰ In rice the identification of dwarf ***Dee Gee Woo Gen*** from a tall rice variety by a Taiwan farmer revolutionized rice breeding. ✰ Using this ***DGWG*** at IRRI during **1965** the wonder rice IR 8 and at Taiwan TN 1 were released
	C.A.Barber and T.S.Venkatraman	✰ Nobilisation in sugarcane (*Saccharum barberi x S. officinarum*) is another monumental work in plant breeding.
1967	Sir Otto Frankel	Coined the term Germplasm
1970	Borlaug, N. E.	Received Nobel Prize for the Green Revolution
1970	C. T. Patel	Developed world's first cotton hybrid (H-4) for commercial cultivation in India.
1970	Berg, Cohen, and Boyer	Introduced the recombinant DNA technology
1974	IBPGR	Establishment of International Board of Plant Genetic Resources
1977	Max Planck Institute of Developmental Biology, Germany	Produced the intergeneric hybrid *Raphanobrassica* (2n=36, RRCC).
1987	Monsanto Company, USA	Developed World's first transgenic cotton in USA.
1991	ICRISAT	Develped world's first pigeonpea hybrid (ICPH 8) for commercial cultivation in India.

Contd...

Table 1–*Contd...*

Year	*Scientist*	*Contribution*
1992	IPGRI	Renamed the IBPGR as International Plant Genetic Resource Institute
1994	"FlavrSavr"	Tomato developed as first genetically modified food produced for the market
1994	Pomato	Produced with fruits by grafting.
1994	Monsanto	Developed the Terminator gene technology.
1995	–	*Bt* corn developed
1996	Monsanto	Roundup Ready genetically modified soybean introduced as Glyphosate resistant. Inhibit a key enzyme ESPS (3 Enol, pyruvynl Shikimic acid-5-Phosphate Synthase) in an amino acid pathway.
1996	Monsanto	BollGard insect resistant transgenic cotton introduced.
1997	Monsanto	YieldGard Corn Borer insect-protected corn is introduced, providing farmers with in-seed insect-protection against the European corn borer. Roundup Ready Canola and Cotton are introduced, providing farmers with in-seed herbicide tolerance to Roundup and other glyphosate-based herbicides.
1998	Monsanto	Roundup Ready Corn tolerance to Roundup and other glyphosate-based herbicides introduced.
1998	Monsanto	Developed the Traitor gene technology
2000	Peter Bayer (Germany) and Ingo Potrykus (Swizerland)	Golden Rice: Genetic manipulation of β-carotene (precursor of Vitamin A) biosynthetic pathway by introgression of psi (phytoene synthase) from Dafodil *(Narcissus pseudoarcissus)* and *crtl* from soil bacterium *Erwinia uredora.* The rice endosperm to accumulate β-carotene (> 1.5 mg/g of dry weight). Reduce colour blindness.
2003	Monsanto	YieldGard Rootworm insect-protected corn is introduced, providing farmers with in-seed insect-protection against the corn rootworm.
2003	-	Iron and Zinc Rice: expression of soybean ferritin gene in indica rice grain endosperm. Help in combating the Anemia.
2002	–	Bt cotton released in India for commercial cultivation
2004		Roundup Ready wheat developed
2005	-	Golden Rice 2: resulting in 23 fold increase in β-carotene content (upto 37mg/g of dry weight)
2013	Thompson and Morgon Co., UK	Sold pre-grafted pomato plants branded as Tomtato.
2013	Bangladesh	Introduced Bt Brinjal for commercial cultivation
2013	Indonesia	Developed the first genetically modified drought resistant Sugarcane.
2014	J. R. Simplot company, USA	Innate Potato: GM potato which will not discolour when peeled and for increased shelflife.
2016	Brazil	Introduce GM virus resistant beans for planting.

- ☆ The discovery of chromosomes as carriers of genes has led to the development of specialized plant breeding methods for chromosome engineering.
- ☆ The totipotency of plant somatic and gametic cells allows regeneration of complete plants from single cells. This, coupled with the development of recombinant DNA technology, has enabled the transfer of desirable genes from any organism into plants. Crop varieties developed in this manner are already in cultivation in several countries.

Table 2: Important Milestones and Development of Plant Breeding in India

Year	*Activities*
1871	The Government of India created the Department of Agriculture
1905	The Imperial Agricultural Research Institute was establish in Pusa, Bihar
1934	The buildings of the institute damaged in earthquake
1936	Shifted to New Delhi
1946	Name was changed Indian Agricultural Research Institute
1901-05	Agricultural Colleges were established at Kanpur, Pune, Sabour, Llyalpur, Coimbatore
1929	Imperial council of Agricultural Research was established
1946	Name was change to Indian Council Agricultural Research
1921	☆ Indian Central Cotton Committee was established ☆ Notable researches on breeding and cultivation of cotton. ☆ Ex: 70 improved varieties of cotton.
1956	Project for intensification of regional research on cotton, oilseeds and millets (PIRRCOM) was initiated to intensify research on these crops – located at 17 different centres throughout the country
1957	All India Coordinated maize improvement project was started with objective of exploiting heterosis
	ICAR initiated coordinated projects for improvement of the other crops
	Total No. of AICRPs : 60
	Total No. of Network Projects: 19 (*Ex:* All India Network Project on Tobacco at Rajamundry)
	Other projects: 10 (*Ex:* Technical Mission on Cotton at Nagpur).
1960	First Agricultural University established at Pantnagar, Nainital, and Uttarkhand.
1961	The first hybrid maize varieties released by the project
1963	Dr. M. S. Swaminathan Introduced the Mexican dwarf varieties in the India
	and a number of high yielding wheat varieties like Kalyan Sona, Sharbathi Sonara
	were developed.
	C.A.Barber and T.S.Venkatraman : Nobilisation in sugarcane (*Saccharum barberi* x *S.*
	officinarum) is another monumental work in plant breeding.
1970	C. T. Patel: Developed world's first cotton hybrid (H-4) for commercial cultivation in India.
1970	Plant Vaiety Protection Act
1995	Bt Cotton work started in India
2002	Bt cotton released in India for commercial cultivation

Aims/Objectives of Plant Breeding

1. **Higher yield:** The ultimate aim of plant breeding is to improve the yield of economic produce. It may be grain yield, fodder yield, fibre yield, tuber yield, cane yield or oil yield depending upon the crop species. Improvement in yield can be achieved either by evolving high yielding varieties or hybrids.
2. **Improved quality:** Quality of produce is another important objective in plant breeding. The quality characters vary from crop to crop. *e.g.* Grain size, colour, milling and backing quality in wheat. Cooking quality in rice, malting quality in barley, colour and size of fruits, nutritive and keeping quality in vegetables, protein content in pulses, oil content in oilseeds, fibre length, strength and fineness in cotton.
3. **Abiotic resistance**: Crop plants also suffer from abiotic factors such as drought, soil salinity, extreme temperatures, heat, wind, cold and frost; breeder has to develop resistant varieties for such environmental conditions.
4. **Biotic resistance:** Crop plants are attacked by various diseases and insects, resulting in considerable yield losses. Genetic resistance is the cheapest and the best method of minimizing such losses. Resistant varieties are developed through the use of resistant donor parents available in the gene pool.
5. **Change in maturity Duration/Earliness:** Earliness is the most desirable character which has several advantages. It requires less crop management period, less insecticidal sprays, permits new crop rotations and often extends the crop area. Development of wheat varieties suitable for late planting has permitted rice-wheat rotation. Thus breeding for early maturing crop varieties or varieties suitable for different dates of planting may be an important objective. Maturity has been reduced from 270 days to 170 days in cotton, from 270 days to 120 days in Pigeon pea, from 360 days to 270 days in sugarcane.
6. **Determinate Growth :** Development of varieties with determinate growth is desirable in crops like mung, pigeon pea (*Cajanus cajan*), cotton (*Gossypium* sp.), *etc.*
7. **Dormancy:** In some crops, seeds germinate even before harvesting in the standing crop if there are rains at the time of maturity, *e.g.*, Groundnut, Green gram, Black gram, Barley and Pea, *etc.* A period of dormancy has to be introduced in these crops to check loss due to germination. In some other cases, however, it may be desirable to remove dormancy.
8. **Desirable Agronomic Characteristics:** It includes plant height, branching, tillering capacity, growth habit, erect or trailing habit *etc.*, is often desirable. For example, dwarfness in cereals is generally associated with lodging resistance and better fertilizer response. Tallness, high tillering and profuse branching are desirable characters in fodder crops.

9. **Elimination of Toxic Substances:** It is essential to develop varieties free from toxic compounds in some crops to make them safe for human consumption. For example, removal of neurotoxin in khesari (*Lathyrus sativus*) which leads to paralysis of lower limbs, Erucic acid from *Brassica* which is harmful for human health, and gossypol from the seed of cotton is necessary to make them fit for human consumption. Removal of such toxic substances would increase the nutritional value of these crops.

10 **Non-shattering characteristics:** The shattering of pods is serious problem in green gram. Hence resistance to shattering is an important objective in green gram.

11. **Synchronous Maturity:** It refers to maturity of a crop species at one time. The character is highly desirable in crops like green gram, cowpea, and cotton where several pickings are required for crop harvest.

12. **Photo and Thermo insensitivity:** Development of varieties insensitive to light and temperature helps in crossing the cultivation boundaries of crop plants. Photo and thermo- insensitive varieties of wheat and rice has permitted their cultivation in new areas. Rice is now cultivated in Punjab, while wheat is a major *rabi* crop in West Bengal.

13. **Wider adaptability:** Adaptability refers to suitability of a variety for general cultivation over a wide range of environmental conditions. Adaptability is an important objective in plant breeding because it helps in stabilizing the crop production over regions and seasons.

14. **Varieties for New Seasons**: Traditionally Maize is a *kharif* crop. But scientists are now able to grow Maize as *rabi* and *zaid* crops. Similarly, mung is grown as a summer crop in addition to the main kharif crop.

Plant Breeding: The Past, Present and Future Scopes

Indian agriculture remained stagnant particularly during early sixties. Long spells of severe drought and serious outbreak of disease in some parts of the country led some futurologists to state that a possible doom in India by the end of the decade. However we achieved a breakthrough in crops such as rice, wheat, pearl millet, jowar and maize.

The indica x japonica cross derivative ADT 27 (cross derivative of GEB 24 x Norin 10) is the first high yielding rice of Tamil Nadu. The identification of *Dee Gee Woo Gen* and release of wonder rice IR 8 (Peta x DGWG) changed the scenario from poverty to problem of plenty.

Likewise identification of dwarfing gene in Japanese wheat variety Norin – 10 by Borlaug and breeding of Mexican dwarf wheat varieties led to the release of wheat varieties like Kalyan Sona in India.

In peal millet, breeding by male sterile line Tift 23A at Tifton, Georgia by Burton and his co-worker and later on its introduction to India led the release of hybrid bajra HB1 to HB4 which increased bajra production many fold. In jowar, breeding of first male sterile line combined kafir 60A and its introduction in to India led to the release of first hybrid sorghum CSH 1 (CK 60A x IS 84) during 1970s.

At present we are in search of alternate source of cytoplasm in almost all crops to breed hybrids with new source of cytoplasm to prevent the possibility of appearance of new pest and diseases.

Scope of Plant Breeding (Future Prospects)

Since the cultivable land is shrinking and there is no scope for increasing the area under cultivation, the only solution to meet the food requirement is by increasing the crop yield through genetic improvement of crop plants. There are two ways by which yield improvement is possible.

1. Enhancing the Productivity of Crops

a) This can be done by the proper management of soil and crops involving suitable agronomic practices and harvesting physical resources.

b) By using high potential crop varieties created by appropriate genetic manipulation of crop plants.

2. Stabilizing the Productivity Achieved

This is done by using crop varieties that are bred especially for wide adaptation or for specific crop zones to offset the ill effects of unfavourable environmental conditions prevailing in the areas.

Advances in molecular biology have sharpened the tools of the breeders, and brighten the prospects of confidence to serve the humanity. The application of biotechnology to field crop has already led to the field testing of genetically modified crop plants. Genetically engineered rice, maize, soybean, cotton, oilseeds rape, sugar beet and alfalfa cultivars are expected to be commercialized before the close of 20th century. Genes from varied organisms may be expected to boost the performance of crops especially with regard to their resistance to biotic and abiotic stresses. In addition, crop plants are likely to be cultivated for recovery of valuable compounds like pharmaceuticals produced by genes introduced into them through genetic engineering. It may be pointed out that in Europe, *hirudin* an anti-thrombin (enzyme in blood plasm which cause the blood clotting by converting fibrinogens into fibrin) protein is already being produced from transgenic *Brassica napus*.

Activities in Plant Breeding

See Figure 1.

Undesirable effects of Plant Breeding

Plant breeding has several useful applications in the improvement of crop plants. However, the main undesirable effects on crop plants.

1. **Genetic erosion:** Disappearance of land races due to introduction of high yielding varieties. *e.g.* Introduction of IR 20 rice led to disappearance of land races of samba rice.

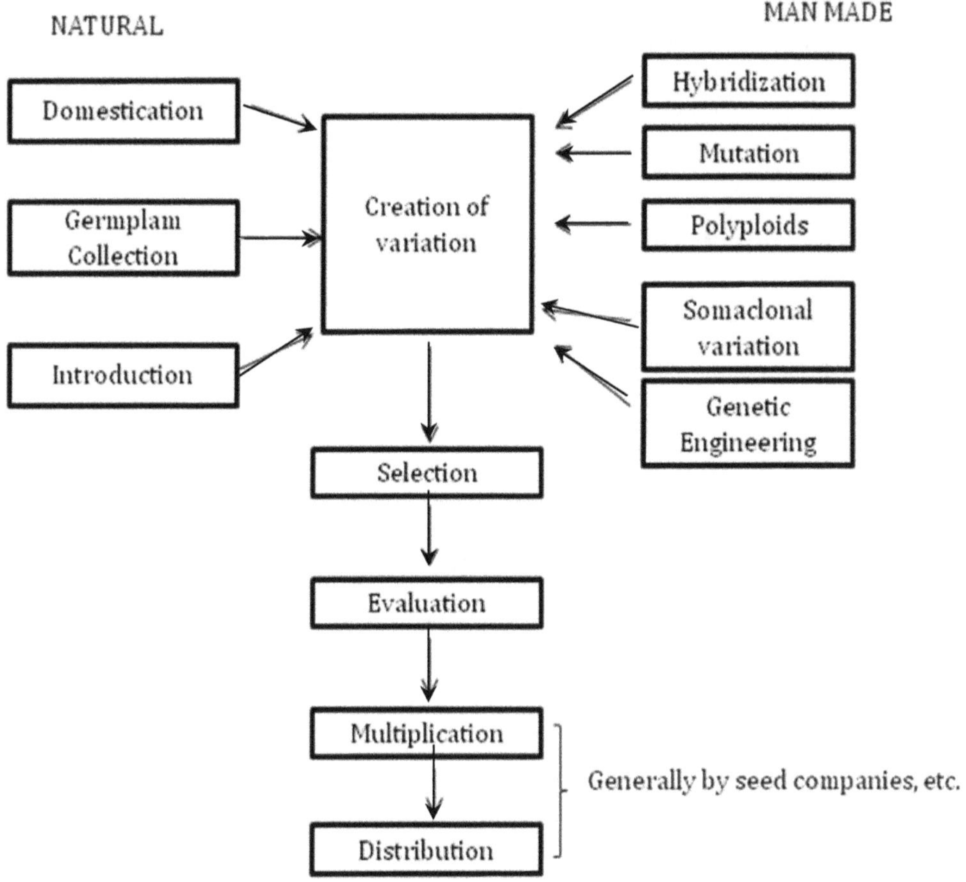

Figure 1: Activities in Plant Breeding.

2. **Narrow genetic base:** Genetic vulnerability to pest and diseases.

Tift 23A	Bajra	Susceptible downy mildew
T cytoplasm	Maize	Susceptible to *Helminthosporium* (Corn leaf blight)

3. **Attainment of yield plateau** - No more further increase in yield.
4. **Danger of Uniformity:** Most of the improved varieties have some common parents in the pedigree which may cause danger of uniformity.
5. **Undesirable combinations:** Sometimes, plant breeding leads to undesirable combinations. The examples of manmade crops having undesirable combination of characters are *Raphano brassica* and *Raparadish* (Karpechenko, 1930) and Pomato.
6. **Increased susceptibility to minor diseases and pests:** Due to emphasis on breeding for resistance to major diseases and insect pests often resulted in

an increased susceptibility to minor diseases and pests. These have gained importance and, in some cases, produced severe epidemics. The epidemic caused by *Botrytis cinerea* (grey mold) in chickpea during 1980-82 Punjab, Haryana. The severe infection by Karnal bunt (*Tilletia* sp.) on some wheat varieties, RTV (Rice Thungro Virus) and infestation of mealy bugs in Bt cotton.

Plant Breeding: Its Relation with other Disciplines

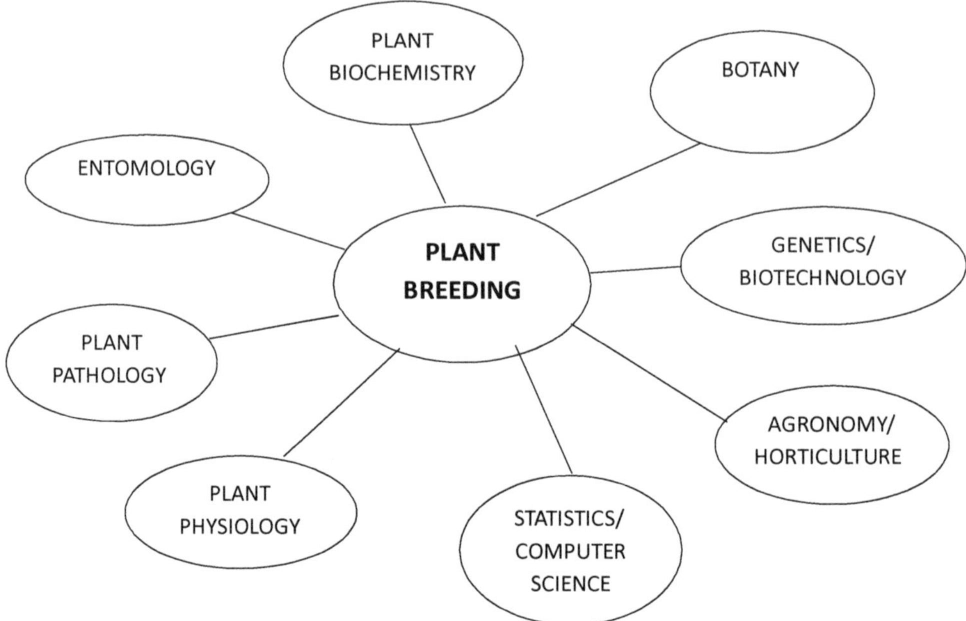

Figure 2: Plant Breeding and its Relationship with other Disciplines.

Genetics/Biotechnology

The plant breeder needs a thorough understanding of the mechanism of heredity in plants as modern plant-breeding methods are based on knowledge of the gene and its inheritance. With the advances in molecular genetics, the knowledge of genes has been extended to the molecular level.

Plant Physiology

Cultivar adaptation is influenced by the response of plants to environmental stresses, such as extremes in temperature, light, soil moisture, and soil nutrients. The plant breeder strives to modify the plant's physiological processes that will enable it to function more efficiently in the environment in which it is grown.

Plant Pathology

Healthy plants are essential for good crop performance. The plant breeder cooperates with the plant pathologist in identification of genes for resistance to plant disease pathogens. Incorporation of genes for resistance to disease into cultivars improves plant performance and reduces the need for chemical disease control.

Entomology

Breeding for insect resistance is an economical and an environmentally sound means for avoiding insect damage while reducing the use of pest control chemicals in field and horticultural crops.

Plant Biochemistry

The inherent nutritional value of a crop cultivar for food or for livestock feed, or for utilization by industry, often may be improved by plant breeding. Examples are texture and flavor in tomato, increased lysine content in feed grains, milling and baking quality in wheat, or fiber fineness and strength in cotton. Molecular genetics has contributed toward a better understanding of the chemical structure and function of the genetic material.

Statistics

The performances of genetically similar strains are compared in the breeding nursery. The breeder needs to be familiar with field plot evaluation techniques that will generate reliable data and statistical procedures to interpret the data accurately. Analytical statistical procedures provide a better understanding of quantitative genetics and its utilization in breeding for improved plant performance.

Computer Science

The computer has become an essential tool for systematic planning of the breeding nursery, recording observations, and rapid analysis and interpretation of the data.

Agronomy/Horticulture

Breeders need to know crops and how to produce them. They should understand the grower's needs in new cultivars of field or horticultural crops in order to evaluate available breeding materials, plan efficient breeding procedures, and direct breeding efforts toward important breeding goals.

A plant breeder cannot be a specialist in all of these fields of plant science. In the practice of plant breeding, the breeder is not working exclusively in any one of them. The task of the plant breeder is to apply the whole of his knowledge and experience with plants toward the development of superior cultivars. If additional information is needed about the inheritance of a plant character, or about a procedure for measuring the comparative tolerance of different cultivars in a particular environment, the breeder may initiate research to study those particular problems. Joint research efforts between the breeder and research specialists in related disciplines to solve common problems are desirable activities to carry out in conjunction with a breeding program.

Salient Achievements of Plant Breeding

- ☆ Directly or indirectly, plant breeding deserves credit for most mans food today, is how Burton (1981) aptly describe the achievements of plant breeding.
- ☆ The yield levels have in general been doubled, though even five to six fold increases have been witnessed during last half a century.
- ☆ The enhanced yield potential of crop variety has indirectly decreased the conversion of forests and grasslands for cultivation of food thus contributing to preservation of bio-diversity.
- ☆ Development of varieties for specific problem areas, *e.g.* drought, water logged *etc.*
- ☆ The discovery of mutation and polyploidy provided for entirely new types of crop plants.
- ☆ The introduction of dwarfing genes in wheat and rice provided in roads to utilize higher levels of chemical fertilizer to attain unprecedented yield levels in the era of green revolution.
- ☆ During 1960's Norman, E. Borlaug, Mexican semi dwarf wheat varieties which paved the way for green revolution in wheat.
- ☆ Dr. M.S. Swaminathan: Introduce and develped a number of high yielding wheat varieties like Kalyan Sona, Sharbathi Sonara.
- ☆ Using this *Dee Gee Woo Gen* at IRRI during 1965 the wonder rice IR 8 was released.
- ☆ Varieties have been developed in crops like soybean and cotton with special plant frame suitable for mechanised harvesting.
- ☆ Maturity period of drops like wheat, rice, cotton, maize *etc.*, has been modified to adopt the new cropping systems.
- ☆ The discovery of *opaque* 2 and floury 2 provided unique lines with superior quality with double the lysine content than normal maize.
- ☆ Nobilisation in sugarcane by C.A. Barber and T.S. Venkatraman.
- ☆ Evaluation of canola type with '0' erucic acid in *Brassica* sps, is an example of enhancing the quality of edible oil.
- ☆ Development of biotic and abiotic stress resistant varieties in majority of the crops.
  - ❒ The epidemics from corn leaf blight in US corn belt, was controlled by genetic improvement of maize varieties.
- ☆ Development of high yielding hybrids/varieties
  - ❒ Hybrid maize: Ganaga safed 2, Deccan, African tall, *etc.*
  - ❒ Jowar : CSH 1, 2, 3, 4, 5, 6, 7, 8, 9, 10, 12, 14, 15R, *etc.*
  - ❒ Bajra: PHB 10, 14, ICTP 8203, WCC 75, *etc.*

- Composites: Manjari, Vikram, Sona, Vijay and Krishna
- Cotton: H 4, DCH 32, NHH 44, *etc.*

Table 3: The Eminent Plant Breeders of India

T.S. Venkatraman	An eminent sugarcane breeder, he transferred thick stem and high sugar contents from tropical noble cane to North Indian Canes. This process is known as noblization of sugarcane
B.P. Pal	An eminent Wheat breeder, developed superior disease resistant N.P. varieties of wheat (NP 700, 800, 114, 775, 4, 710, 770, 783 and 824).
M.S. Swaminathan	Responsible for green revolution in India, developed high yielding varieties of Wheat and Rice.
Pushkarnath	Famous potato breeder
N.G.P. Rao	An eminent sorghum breeder
K. Ramaiah	A renowned rice breeder
Ram Dhan Singh	Famous wheat breeder
D.S. Athwal	Famous pearl millet breeder
Bosisen	An eminent maize breeder
Dharampal Singh	An eminent oil- seed breeder
C.T. Patel	Famous cotton breeder who developed world's first commercial intraspecific cotton hybrid in 1970 (H4).
V. Santhanam	Famous cotton breeder

Agencies Involved in Plant Breeding Activities

Table 4: The Members of the CGIAR (Consortium of International Agricultural Research Centers)

Year of Establish ment	*Acronym*	*Active CGIAR Centers*	*Headquarters Location*
1960	IRRI	International Rice Research Institute	Manila, Philippines
1966	CIMMYT	International Maize and Wheat Improvement Center	Mexico State, Mexico
1967	IITA	International Institute of Tropical Agriculture	Ibadan, Nigeria
1968	CIAT	International Center for Tropical Agriculture	Cali, Colombia
1971	WARDA	West Africa Rice Development Association	Monrovia, Liberia
1971	CIP	International Potato Center	Lima, Peru
1972	ICRISAT	International Crops Research Institute for the Semi-Arid Tropics	Hyderabad (Patancheru), India
1974	IBPGR	International Borard of Plant Genetic Research	Rome, Italy
1974	BI	Biodiversity International (Research for develop ment in agriculture and forest biodiversity)	Maccarese, Rome, Italy
1975	ICLARM	World Fish Center (International Center for Living Aquatic Resources Management)	Penang, Malaysia

Contd...

Table 4–*Contd...*

Year of Establish ment	*Acronym*	*Active CGIAR Centers*	*Headquarters Location*
1975	IFPRI	International Food Policy Research Institute	Washington, D.C., United States
1976	ICARDA	International Center for Agricultural Research in the Dry Areas	Aleppo, Syria
1978	ICRAF	World Agroforestry Centre (International Centre for Research in Agroforestry)	Nairobi, Kenya
1984	IWMI	International Water Management Institute	Battaramulla, Sri Lanka
1993	CIFOR	Center for International Forestry Research	Bogor, Indonesia
1994	ILRI	International Livestock Research Institute	Nairobi, Kenya

2

Modes of Reproduction

Mode of reproduction determines the genetic constitution of crop plants, that is, whether the plants are normally homozygous or heterozygous. This, in turn, determines the goal of a breeding programme. If the crop plants are naturally homozygous, *e.g.*, as in self- pollinators like wheat, a homozygous line would be desirable as a variety. But if the plants are heterozygous naturally, *e.g.*, as in cross-pollinators like maize, a heterozygous population has to be developed as a variety. Consequently, the breeding methods have to be vastly different for the two groups of crop plants. Knowledge of the mode of reproduction of crop plants is also important for making artificial hybrids. Production of hybrids between diverse and desirable parents is the basis for almost all the modern breeding programmes.

The modes of reproduction in crop plants may be broadly grouped into two categories, *asexual* and *sexual.*

SEXUAL REPRODUCTION

Sexual reproduction involves fusion of male and female gametes to form a zygote, which develops in to an embryo. In crop plants, male and female gametes are produced in specialised structures known as flowers.

Flower Structure (Figure 3)

- **Sepals:** Modified leaves that enclose and protect the flower before it opens
- **Petals:** Brightly colored parts that aid in attracting pollinators
- **Stamens:** Includes the anther and filament, the male reproductive organs
- **Anther:** Where the pollen is produced
- **Filament:** The stalk which the anther is at the end of

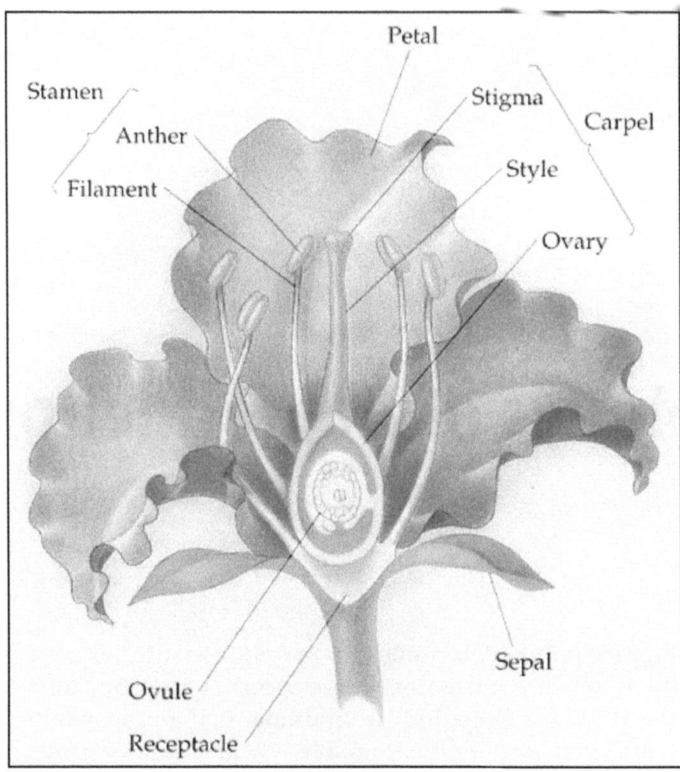

Figure 3: A Typical Structure of Flower.

- **Carpel:** Includes the stigma, style, and ovary, the female reproductive organs
- **Stigma:** Sticky end that receives the pollen
- **Style:** Long stem that connects to the ovary
- **Ovary:** Organ where fertilization takes place

* Only angiosperms produce flowers, gymnosperms do not have this structure.

Types of Flowers

A flower containing both stamens and pistil is a perfect or hermaphrodite flower. If it contains stamens, but not pistil, it is known as staminate, while a pistillate flower contains pistil, but not stamens. Staminate and pistillate flowers occur on the same plant in a monoecious species, such as maize, colocasia, castor (*Ricinus communis*), coconut, *etc.* But in dioecious species, staminate and pistillate flowers occur on different plants, *e.g.*, papaya, datepalm (*Phoenix dactylifera*), pistachio (*Pistacia vera*), gynoecious (cucumber), gynomonoecious (members of compositae family), andromonoecious (muskmelon, tea, passion flower), trimonoecious (papaya) *etc.* In "crop plants, meiotic division of specific cells in stamen and pistil yields microspores and megaspores, respectively. This is followed by mitotic division of the spore nuclei to produce gametes; the male and female gametes are produced in microspores and megaspores, respectively.

Table 5: Types of Flowers Based on Individual Flowers and Individual Plants

Sexuality	*Phenotype*	*Description*
Individual flowers		
Hermaphrodite	⚥	Bisexual flower with stamens and pistil
Female	♀	Unisexual pistillate flower
Male	♂	Unisexual starminate flower
Individual plants		
Hermaphrodite	⚥	Only hermaphrodite flowers
Monoecious	♀♂	Female and male flowers on same plant
Dioecious	♀♂	Female and male flowers on different plant
Gynomonoecious	⚥ ♀	Hermaphrodite and female flowers on same same plant
Andromonoecious	⚥ ♂	Hermaphrodite and male flowers on same same plant
Trimonoecious	⚥ ♀♂	Hermaphrodite, female and male flowers on same same plant

Sporogenesis

Productions of microspores and megaspores is known as **sporogenesis.** Microspores are produced in anthers (microsporogenesis), while **megaspores** are produced in ovules (megasporogenesis).

Microsporogenesis

Each anther has four pollen sacs, which contain numerous pollen mother cells (PMCs). Each PMC undergoes meiosis to produce four haploid cells or microspores. This process is known as microsporogenesis. The microspores mature into pollen grains mainly by a thickening of their walls.

Megasporogenesis

Megasporogenesis occurs in ovules, which are present inside the ovary. A single cell in each ovule differentiates into a megaspore mother cell. The megaspore mother cell undergoes meiosis to produce four haploid megaspores. Three of the megaspores degenerate leaving one functional megaspore per ovule (*Figure* 4). This completes megasporogenesis.

Gametogenesis

The production of male and female gametes in the microspores and the megaspores, respectively, is known as gametogenesis.

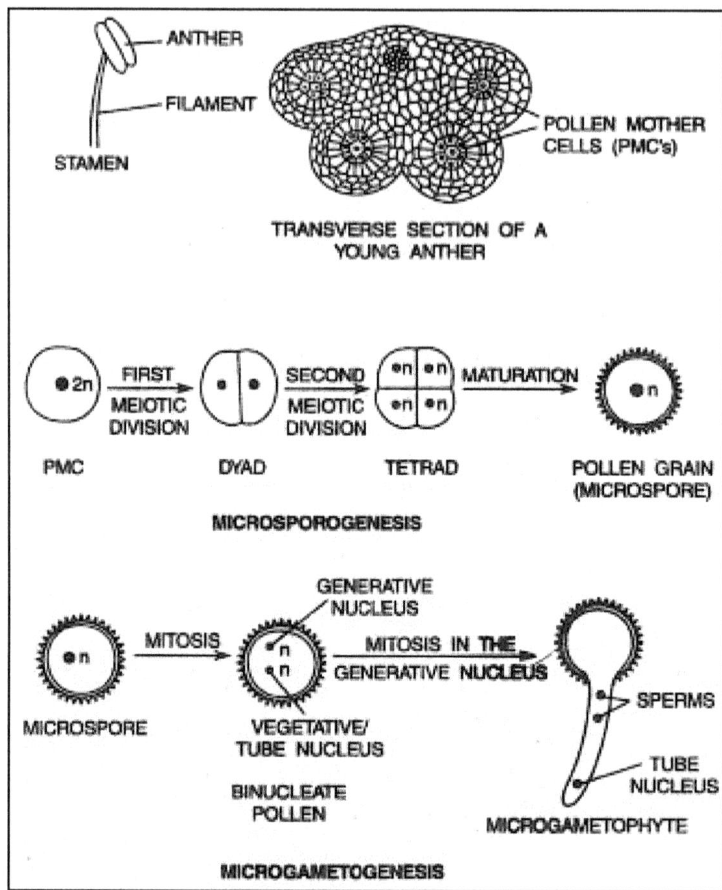

Figure 4: Microsporogenesis and Microgametogenesis (A generalized scheme).

Microgametogenesis

This refers to the production of male gamete or **sperm.** During the maturation of pollen, the microspore nucleus divides mitotically to produce a generative and a vegetative or tube **nucleus.** The pollen is generally released in this binucleate stage. When the pollen lands onto the stigma of a flower, it is known as pollination. Shortly after pollination, the pollen germinates. The pollen tube enters the stigma and grows through the style. The generative nucleus now undergoes a mitotic division to produce two male gametes or sperms. The pollen, along with the pollen tube, is known as microgametophyte. The pollen tube finally enters the ovule through a small pore, micropyle, and discharges the two sperms into the embryo sac.

Megagametogenesis

The nucleus of a functional megaspore divides mitotically to produce four or more nuclei. The exact number of nuclei and their arrangement vary considerably from one species to another. In most of the crop plants, megaspore nucleus

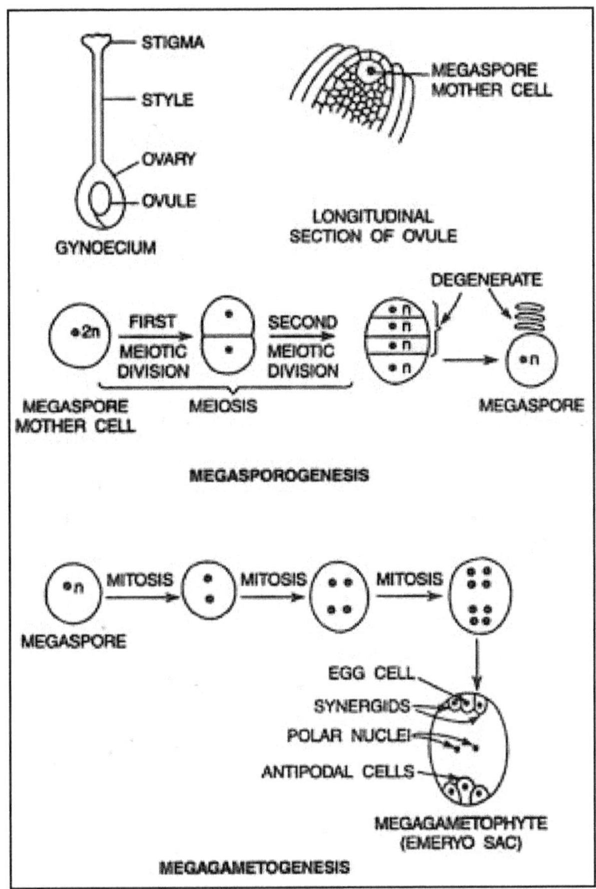

Figure 5: Megasporogenesis and Megagametogenesis (Generalized scheme).

undergoes three mitotic divisions to produce eight nuclei. Three of these nuclei move to one pole and produce a central egg cell and two synergid cells; one synergid is situated on either side of the egg cell. Another three nuclei migrate to the opposite pole to give rise to antipodal cells. The two nuclei remaining in the centre, the polar nuclei, fuse to form a secondary nucleus. The megaspore thus develops into a mature megagametophyte or embryo sac. The development of embryo sac from a megaspore is known as megagametogenesis. The embryo sac generally contains one egg cell, two synergids, three antipodal cells (all haploid), and one diploid secondary nucleus.

Fertilization

The process of fusion of the male and female gametes to form a zygote is called fertilization.

1. Pollen lands on a stigma.
2. A pollen tube grows from pollen grain, down the style, through the micropyle into the ovule.

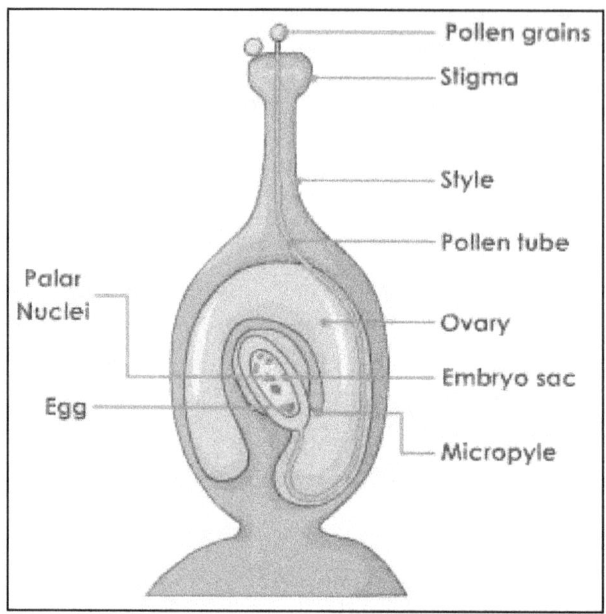

Figure 6: Fertilization in Flowering Plant.

3. 2 sperm nuclei discharged in the embryo sac results in double fertilization.
4. The zygote develops in the embryo.
5. The fertilized central cell becomes the endosperm.
6. The ovule develops into a seed.
7. The ovary develops into a fruit.

Significance of Sexual Reproduction

Sexual reproduction makes it possible to combine genes from two parents into a single hybrid plant. Recombination of these genes produces a large number of genotypes. This is an essential step in creating variation through hybridization. Almost the entire plant breeding is based on sexual reproduction. Even in asexually reproducing species, sexual reproduction, if it occurs, is used to advantage, *e.g.*, in sugarcane, potato, sweet potato *etc.*

ASEXUAL REPRODUCTION

Asexual reproduction does not involve fusion of male and female gamentes. New plants may develop from vegetative parts of the plant (*vegetative reproduction)* or may arise from embryos that develop without fertilization (apomixis).

Natural Vegetative Reproduction

In nature, a new plant develops from a portion of the plant body. This may occur through modified underground and sub-aerial stems, and through bulbills.

Underground Stems

The underground modifications of stem generally serve as storage organs and contain many buds. These buds develop into shoots and produce plants after rooting. Examples of such modifications are given below.

- Tuber: Potato, Asparagus, *etc.*
- Bulb: Onion, Garlic, Tuberose, *etc.*
- Rhizome: Ginger, turmeric (*Curcuma domestica*), *etc.*
- Corm: Bunda (*Colocasia antiquorum*), Arwi (*Colocasia esculenta*), Gladiolus, *etc.*

Sub-aerial Stems

These modifications include runner, stolon, sucker *etc.* Sub-aerial stems are used for the propagation of mint, date plam, *etc.*

Bulbils

Bulbils are modified flowers that develop into plants directly without formation of seeds. These are vegetative bodies; their development does not involve fertilization and seed formation. The lower flowers in the inflorescence of garlic naturally develop into bulbils. Scientists are trying to induce bulbil development in plantation crops by culturing young inflorescence on tissue culture media; it has been successfully done in the case of cardamom.

Artificial Vegetative Reproduction

It is commonly used for the propagation of many crop species, although it may not occur naturally in those species.

- *Stem cuttings:* Sugarcane, Grapes, Roses, *etc.*
- *Root cuttings:* Curry leaf.
- *Leaf cutting:* Begonia.
- *Layering:* Guava, Gooseberry, Apple, Jasmine, *etc.*
- *Air layering:* Thuja.
- *Budding:* Rose, Ber, *etc.*
- *Grafting:* Sapota, Mango, *etc.*
- *Tissue culture*: Banana, Gerbera, Cornation *etc.*

Stem cuttings are commercially used for the propagation of sugarcane, grapes, roses, *etc.* Layering, budding, grafting and gootee are in common use for the propagation of fruit trees and ornamental shrubs. Techniques are available for vegetative multiplication through tissue culture in case of many plant species, and attempts are being made to develop the techniques for many others. In many of these species sexual reproduction occurs naturally but for certain reasons vegetative reproduction is more desirable.

Significance of Vegetative Reproduction

Vegetatively reproducing species offer unique possibilities in breeding. A desirable plant may be used as a variety directly regardless of whether it is homozygous or heterozygous. Further, mutant buds, branches or seedlings, if desirable, can be multiplied and directly used as varieties.

Advantages and Disadvantages of Natural Vegetative Propagation

Advantages

- No need of external agencies for pollination, *e.g.* insects, wind, *etc.*
- Food is usually present in the vegetative structures; a rapid development of buds into daughter plants can take place.
- New plants resemble parent plant in every way.
- Involves only one parent.

Disadvantages

- Lack of dispersal mechanism may lead to overcrowding.
- New plants are less varied
- New plants may be less adaptable to changes in environmental condition.

APOMIXIS

Apomixis, derived from two Greek word "APO" (*away from*) and "mixis" (*act of mixing or mingling*). It refers to the occurrence of an asexual reproductive process in the place of normal sexual processes involving reduction division and fertilization. In other words apomixis is a type of reproduction in which sexual organs of related structures take part but seeds are formed without union of gametes. Seeds formed in this way are vegetative in origin. When apomixis is the only method of reproduction in a plant species, it is known as obligate apomixis. On the other hand, if gametic and apomictic reproduction occurs in the same plant, it is known as facultative apomixis. The first discovery of this phenomenon is credited to Leuwenhock as early as 1719 in citrus seeds.

Apomixis is widely distributed among higher plants. More than 300 species belonging to 35 families are apomictic. It is most common in Gramineae, Compositae, Rosaceae and Rutaceae. Among the major cereals maize, wheat and pearl millet have apomictic relatives.

Long back, Winkler (1908) defined apomixis as "the substitution for sexual reproduction or another asexual reproductive process that does not involve nuclear or cellular fusion (*i.e.* fertilization)". Stebbins (1914) and later Nygren (1954) presented an excellent review on apomixis in angiosperms.

Types of Apomixis

Mainly four types of apomixis phenomenon are suggested by Maheshwari (1954).

1. Recurrent Apomixis

An embryo sac develops from the MMC or megaspore mother cell (archesporial cell) where meiosis is disturbed (sporogenesis failed) or from some adjoining cell (in that case MMC disintegrates). Consequently, the egg-cell is diploid. The embryo subsequently develops directly from the diploid egg-cell (MMC) *i.e.,* without fertilization. **Somatic apospory, diploid parthenogenesis and diploid apogamy are recurrent apomixis.** However, diploid parthenogenesis/apogamy occurs only in aposporic (somatic) embryo-sacs. Therefore, it is the somatic or diploid aposory that constitutes the recurrent apomixis. Such apomixis occurs in some species of *Crepis, Taraxacum, Poa* (blue grass), and *Allium* (onion) without the stimulus of pollination. In *Malus* (apple) and *Rudbeckia* pollination appears to be necessary, either to stimulate embryo development or to produce a viable endosperm.

2. Non-recurrent Apomixis

An embryo arises directly from normal egg-cell (n) without fertilization. Since an egg-cell is haploid, the resulting embryo will also be haploid. Haploid parthenogenesis and haploid apogamy, and androgamy fall in this category. Such types of apomixis are of rare occurrence. They do not perpetuate and are primarily of genetic interest as in corn.

3. Adventive Embryony

Embryos arise from a cell or a group of cells either in the nucellus or in the integuments, *e.g.* in oranges and roses. Since it takes place outside the embryo sac, it is not grouped with recurrent apomixis, though this is regenerated with the accuracy. In addition to such embryos, regular embryo within the embryo sac may also develop simultaneously, thus giving rise to poly-embryony condition, as in *Citrus, Opuntia.*

Development of Apomictic Embryo Sac

1. Apospory

It involves the development of embryo sac either from the archesporial cell (the primitive cell, or group of cells, that give rise to the cells from which spores are derived) or from the nucellus, or from other cell. It is of two types:

(i) *Generative or haploid apospory*: If the embryo sac develops from one of the megaspores (n), the process is called generative or haploid apospory. Since it cannot regenerate, as it is haploid and fertilization fails, the process gives rise to non-recurrent apomicts.

(ii) *Somatic or diploid apospory*: When diploid embryo sac is formed from nucellus or other cells, the process is termed as somatic or diploid asopory. Since it regenerates without fertilization, it is recurrent. *e.g. Hierarceum, Malus, Creps, Ranunculus, Orchids, etc.*

Development of Apomictic Embryo

1. Parthenogenesis

This refers to the development of embryo from egg-cell without fertilization, *e.g.* in some cases in corn, wheat, tobacco. This is also of two kinds:

(i) ***Haploid parthenogenesis*:** The embryo develops from egg-cell without fertilization in a haploid embryo-sac produced by generative apospory. It is non-recurrent in nature.

(ii) ***Diploid parthenogenesis*:** The embryo develops from egg-cell without fertilization in a diploid embryo-sac arising from somatic apospory. It is recurrent type.

2. Apogamy

This is related to the development of embryo not from the egg-cell but from any one of the synergid or antipodal cells within the embryo sac, without fertilization. This is haploid or diploid. In the haploid apogamy, the embryo arises from any cell other than the egg-cell without fertilization in haploid embryo -sac formed by generative apospory. By virtue of its haploid nature, it is also non-recurrent apomixis. Whereas in case of diploid apogamy embryo develops from any cell other than the egg-cell without fertilization in a diploid embryo-sac developed by somatic apospory. It is recurrent type.

3. Androgamy

It is the development of embryo neither from egg cell nor from synergids or antipodals, but from one of the male gametes itself, inside or outside the embryo-sac. Since it is haploid, it is non-recurrent apomixis. *e.g. Nicotiana*, Datura, Rice, *etc.*

In **parthenocarpy**, seedless fruits are formed from ovary without fertilization. Normally, fertilization stimulates the ovary to become enlarged and form fruit. But in case of parthenocarpy, such stimulation may be received even from incompatible pollination. *e.g.* Banana.

Advantages of Apomixis in Plant Breeding

1. Rapid multiplication of genetically uniform individuals can be achieved without risk of segregation.
2. Heterosis or hybrid vigour can permanently be fixed in crop plants, thus no problem for recurring seed production of F_1 hybrids.
3. Efficient exploitation of maternal effect, if present, is possible from generation to generation.
4. Apomictic hybrid seed plots need not be grown in isolation.

Detection of Apomixis

☆ Failure to obtain F_1 plants from a cross (using apomictic female).

☆ Uniform progeny from a heterozygous plant or from open pollinated sps.

- ☆ Production of maternal phenotype in the progeny from crosses.
- ☆ Formation of irregular embryo sacs.
- ☆ Multiple seedlings per seed, flowers with two or more fused ovaries.
- ☆ Multiple stigmas and multiple ovules per floret.

Significance of Apomixis

Apomixis is a nuisance when the breeder desires to obtain sexual progeny, *i.e.*, selfs or hybrids. But it is of great help when the breeder desires to maintain varieties. Thus in breeding of apomictic species, the breeder has to avoid apomictic progeny when he is making crosses or producing inbred lines. But once a desirable genotype has been selected, it can be multiplied and maintained through apomictic progeny. This would keep the genotype of a variety intact. Asexually reproducing crop species are highly heterozygous and show severe inbreeding depression. Therefore, breeding methods in such species must avoid inbreeding.

3

Modes of Pollination

Pollination refers to the transfer of pollen grains from anthers to the receptive surface of stigmas. Pollen from an anther may fall on to the stigma of the same flower leading to self- pollination or Autogamy. When pollen from flowers of one plant is transmitted to the stigmas of flowers of another plant, it is known as cross-pollination or allogamy. A third situation, geitonogamy, results when pollen from a flower of one plant falls on the stigmas of other flowers of the same plant, *e.g.*, in maize. The genetic consequences of geitonogamy are the same as those of autogamy provided the different branches of the same plant are genetically identical.

SELF-POLLINATION

Many cultivated plant species reproduce by self-pollination. Self-pollination species are believed to have originated from cross-pollinated ancestors. These species, as a rule, must have hermaphrodite flowers. But in most of these species, self -pollination is not complete and cross -pollination may occur up to 5 per cent. The degree of cross-pollination in self- pollinated species is affected by several factors, *e.g.*, variety, environmental conditions like temperature and humidity and location.

Mechanisms Promoting Self-pollination/Contrivances of Self-pollination

The various mechanisms that promote self- pollination are generally more efficient than those promoting cross -pollination. These mechanisms are listed below.

1. **Cleistogamy (Gr. Kleisto-closed)/Automatic self-pollination (Closed marriage):** In this case, bisexual flowers do not open at all. This ensures complete self- pollination since foreign pollen cannot reach the stigma of a closed flower. Cleistogamy occurs in some varieties of groundnut,

peas, beans and in a number of other grasses. In cleistogamous flowers the pollengrains germinate inside the pollensacs. The pollen tubes pierce out through the antherwall and then enter the stigma of the same flower.

2. **Chasmogamy (Open marriage):** In some species, the bisexual flowers open, but only after pollination has taken place. This occurs in many cereals, such as, wheat, barley, rice and oats. Since the flower does open, some cross-pollination may occur.
3. In crops like tomato and brinjal, the stigmas are closely surrounded by anthers. Pollination generally occurs after the flowers open. But the position of anthers in relation to stigmas ensures self- pollination.
4. In some species, flowers open but the stamens and the sigma are hidden by other floral organs. In several legumes, *e.g.*, pea, mung, urd, soybean and gram the stamens and the stigma are enclosed by the two petals forming a keel.
5. In a few species, stigmas become receptive and elongate through staminal columns. This ensures predominant self -pollination.
6. **Homogamy:** The maturation of stamens and carpels at the same time.

Consequences of Self-pollination

- Self- pollination maintains the parity of the race from generation to generation.
- Self- pollination leads to a very rapid increase in homozygosity. Therefore, populations of self- pollinated species are highly homozygous, self-pollinated species do not show inbreeding depression, but may exhibit considerable heterosis. Therefore, the aim of breeding methods generally is to develop homozygous varieties.

CROSS-POLLINATION

In cross- pollinating species, the transfer of pollen from a flower to the stigmas of the others may be brought about by wind (*anemophily*). Many of the crop plants are naturally cross-pollinated. In many species, a small amount (up to 5- 10 per cent) of selfing may occur.

Mechanisms Promoting Cross-pollination

There are several mechanisms that facilitate cross-pollination; these mechanisms are described briefly.

1. **Dicliny/unisexuality:** It is a condition in which the flowers are either staminate (male) or pistillate (female).
   a) **Monoecy.** Staminate and pistillate flowers occur in the same plant, either in the same inflorescene, *e.g.*, castor, mango and coconut, or in separate inflorescences, maize, chestnut, strawberries, rubber, grapes and cassava.

   b) **Dioecy.** The male and female flowers are present on different plants, *i.e.*, the plants in such species are either male or female, *e.g.*, papaya, date, hemp, asparagus, and spinach. In general, the sex is governed by a single gene, *e.g.*, asparagus and papaya. In some cases, there are hermaphrodite plants in addition to males and females, and a number of intermediate forms may also occur.

2. **Dichogamy:** (Gr: *dicha – in two)* stamens and pistils of hermaphrodite flowers may mature at different times facilitating cross - pollination.
   a) **Protogyny :** (Gr: Protos-first, gyne-female) in crop species like bajra, pistils mature before stamens.
   b) **Protandry:** (Gr: Protos-first, andros-male) in crops like maize, sunflower, allium, ocimum and sugarbeets, stamens mature before pistils.
3. **Herkogamy**: (Gr: herkos-afence or barrier) in certain flowers there is a certain physical barrier which reduce self-pollination. In Lucerne or alfalfa, stigmas are covered with a waxy film. The stigma does not become receptive until this waxy film is broken. The waxy membrane is broken by the visit of honey bees which also effect cross- pollination.

   e.g: Calotropis, Clerdendron, Iris *etc.*
4. A combination of two or more of the above mechanisms may occur in some specie. This improves the efficiency of the system in promoting cross-pollination. For example, Maize exhibits both monoecy and protandry.
5. **Self-Incompatibility/Self-sterility**. It refers to the failure of pollen from a flower to fertilize the same flower or other flowers on the same plant. Self- incompatibility is of two types: sporophytic and gametophytic. In both the cases, flowers do not set seed on selfing. Self- incompatibility is common in several species of *Brassica,* some species of *Nicotiana,* radish, rye and many grasses. It is highly effective in preventing self- pollination.
6. **Male Sterility**. Male sterility refers to the absence of functional pollen grains in otherwise hermaphrodite flowers. Male sterility is not common is natural populations. But it is of great value in experimental populations, particularly in the production of hybrid seed. Male sterility is of two types: genetic and cytoplasmic. Cytoplasmic male sterility is termed Cytoplasmic -Genetic when restorer genes are known. In view of the importance of self- incompatibility and male sterility, a more detailed discussion on them follows later.
7. **Heterostyly:** (Gr: heteros – different) some flowers have short styles and long stamens while other have long styles and short stamens.

   e.g. Primula, Jasmine, Oxalis *etc.*

Genetic Consequences of Cross-pollination

☆ Cross- pollination preserves and promotes heterozygosity in a population. Cross-pollinatedspecies are highly heterozygous and show mild to severe inbreeding depression and a considerable amount of heterosis. The breeding methods in such species aim at improving the crop species without reducing heterozygosity to an appreciable degree. Usually, hybrid or synthetic varieties are the aim of breeder wherever the seed production of such varieties is economically feasible.

☆ Cross-pollination results in agenetic variability, which may be advantageous or disadvantageous.

Table 6: Difference between Self and Cross-pollinated Crops

Sl.No.	*Self-pollination*	*Cross-pollination*
1.	When pollen grains from a flower are carried to the stigma of the same flower or on the other flower of the same plant, the pheno menon is called self pollination.	When the pollen grains from a flower are carried to the stigma of the flower on other plant, the phenomenon is called cross pollination.
2.	Also called Autogamy.	Allogamy.
3.	Flowers exhibiting self pollination sometimes do not need any pollinating agent.	Flowers exhibiting cross pollination needsome pollinating agents.
4.	From the genetical as well as quality point of view Self pollination is less preferable.	From the genetical as well as quality point of view Cross-pollination is better and more preferable than self-pollination.
5.	Seen in Some legumes, orchids *etc.*	Most flowers.
6.	Smaller flowers.	Plant differences: Brightly colored petals, scent, long stamens and pistils.
7.	More uniform progeny; less energy must be spent on attracting pollinators. Allows plant to spread further.	More variety in species.

Table 7: Classification of Crop Plants Based on Mode of Pollination and Mode of Reproduction

Mode of Pollination	*Examples of Crop Plants*
Self Pollinated Crops	Rice, Wheat, Barley, Oats, Chickpea, Pea, Cowpea, Lentil, Green gram, Black gram, Soybean, Common bean, Moth bean, Linseed, Sesame, Khesari, Sunhemp, Chillies, Brinjal, Tomato, Okra, Peanut, Potato, *etc.*
Cross-pollinated Crops	Corn, Pearlmillet, Rye, Alfalfa, Radish, Cabbage, Sunflower, Sugarbeet, Castor, Red clover, White clover, Safflower, Spinach, Onion, Garlic, Turnip, Squash, Muskmelon, Watermelon, Cucumber, Pumpkin, Kenaf, Oilpalm, Carrot, Coconut, Papaya, Sugarcane, Coffee, Cocoa, Tea, Apple, Pears, Peaches, Cherries, grapes, Almond, Strawberries, Pine apple, Banana, Cashew, Irish, Cassava, Taro, Rubber, *etc.*
Often Cross Pollinated Crops	Sorghum, Cotton, Triticale, Pigeonpea, Tobacco *etc.*

OFTEN CROSS-POLLINATED SPECIES

In many crop plants, cross -pollination often exceeds 5 per cent and may reach 30 per cent. Such species are generally known as often cross -pollinated species, *e.g.*, jowar, cotton, arhar, safflower *etc.* The genetic architecture of such crops is intermediate between self- pollinated and cross-pollinatedspecies. Consequently, in such species breeding methods suitable for both of them may be profitably applied. But often hybrid varieties are superior to others.

Pollinating Agents

- **Abiotic Pollination**: Anemophily- air; Hydrophily- H_2O.
- **Biotic Pollination:** Zoophily – animals; Chiropteriphily – bats; Ornithophily- birds; Malacophily - snails and slugs; Myrmecophily – ants; Entomophily – insects; Anthropophily – man, *etc.*

4

Self-Incompatibility

Self-incompatibility or self-sterility and male sterility are the two mechanisms which encourage cross pollination. More than 300 species belonging to 20 families of angiosperms show self incompatibility

Definitions

In self-incompatible plants, the flowers will produce functional or viable pollen grains which fail to fertilize the same flower or any other flower of the same plant.

OR

Ability of plant to reject its own pollen.

Reasons

a) Self-incompatible pollen grain may fail to germinate on the stigmatic surface.

b) Some may germinate but fails to penetrate the stigmatic surface

c) Some pollen grains may produce pollen tube which enters through stigmatic surface but its growth will be too slow. By the time the pollen tube enters the ovule the flower will drop.

d) Some time fertilization is effected but embryo degenerates early.

Self-incompatibility is appeared to be due to biochemical reaction, but precise nature of these reactions is not clearly understood.

Classification of Self-incompatibility

According to Lewis (1954) the self-incompatibility is classified as follows:

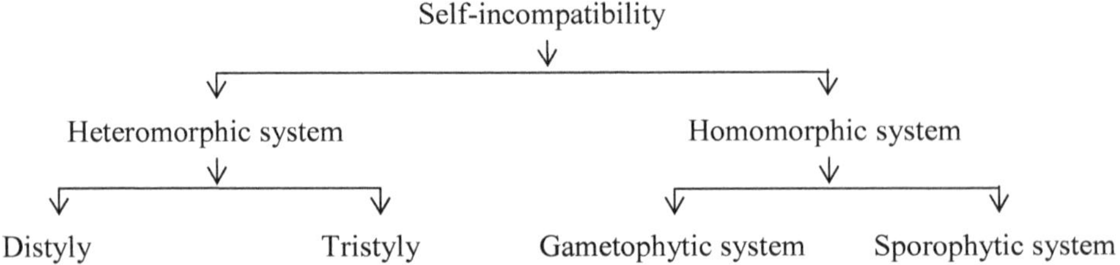

Heteromorphic System

In this case there will be difference in the morphology of the flowers. For example in Primula there are two types of flowers namely PIN and THRUM.

PIN flowers have long style and short stamens, while THRUM flowers have short style and long stamens. This type of difference is known as Distyly.

TRISTYLY is known in some plants like *Lythrum*. In this case the style of the flower may be short, long or medium length.

In case of distyly the only compatible mating is between PIN and THRUM. The relative position of anthers is determined by single gene *S/s*. The recessive ss produce PIN and heterozygote *Ss* produces THRUM. Homozygous dominant *SS* is lethal and do not exist. The incompatibility reaction of pollen is determined by the genotype of the plant producing them. Allele *S* is dominant over *s*. This system is also known as heteromorphic - sporophytic system. Pollen grains produced by PIN flowers will be all *s* in genotype as well as in incompatibility reaction. Whereas THRUM flowers will produce two types of gametes *S* and *s* but all of them would be *S* phenotypically. The mating between PIN and THRUM would produce *Ss* and *ss* progeny in equal frequencies. This system is of little importance in crop plants. It occurs in sweet potato and buck wheat.

Table 8: The Mating between PIN and THRUM (Distyly)

Mating		*Progeny*	
Phenotype	*Genotype*	*Genotype*	*Phenotype*
Pin x Pin	*ss* x *ss*	Incom. mating	–
Pin x Thrum	*ss* x *Ss*	1 *ss* : *Ss*	1 Pin:1 Thrum
Thrum x Pin	*Ss* x *ss*	1 *Ss* : *ss*	1 Thrum:1 Pin
Thrum x Thrum	*Ss* x *Ss*	Incom. mating	–

Homomorphic System

Here the incompatibility is not associated with morphological difference among flower. The incompatibility reaction of pollen may be controlled by the genotype

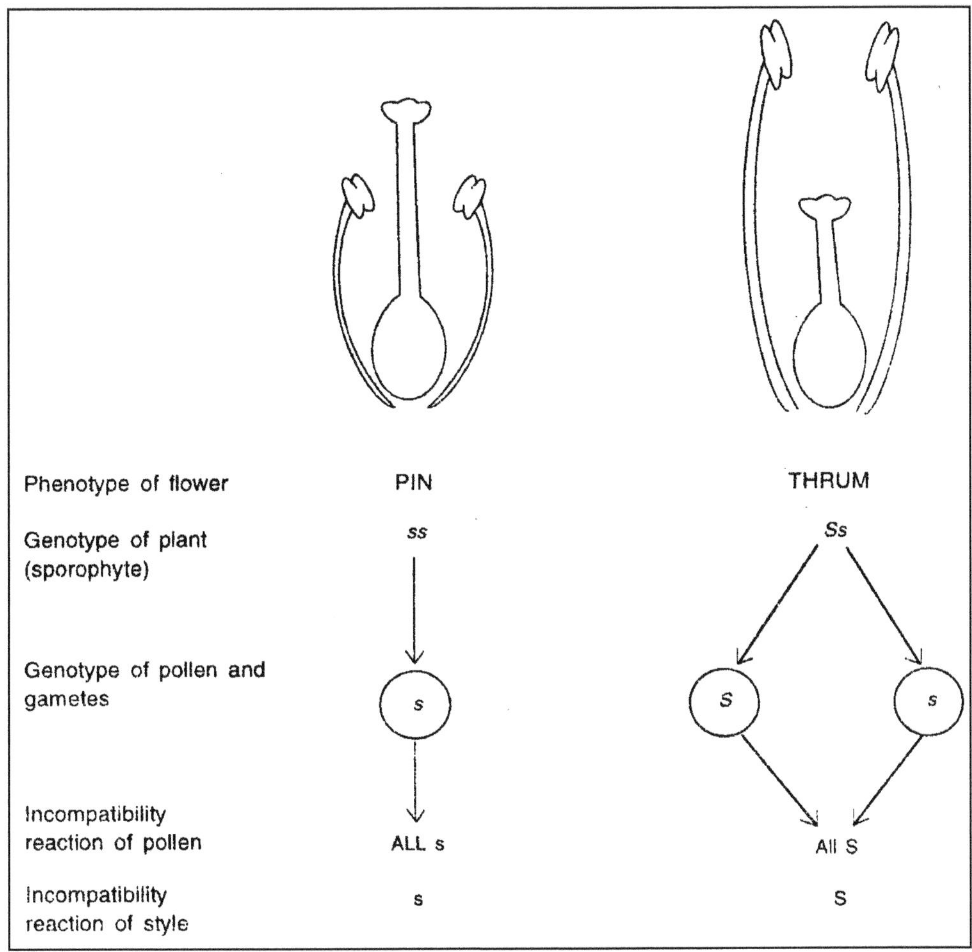

Figure 7: The Mating between PIN and THRUM.

of the plant on which it is produced (Sporophytic control) or by its own genotype (Gametophytic control).

Gametophytic System

First discovered by East and Mangelsdorf in 1925 in *Nicotiana sanderae*. Here the incompatible reaction of pollen is determined by its own genotype and not by the genotype of the plant on which pollen is produced. Generally the incompatibility reaction is determined by a single gene having multiple allele. *e.g. Trifolium, Nicotiana, Lycopersicon, Solanum,* Petunia. Here Codominance is assumed.

Sporophytic System

Here also the self incompatibility is governed by a single gene S with multiple alleles. More than 30 alleles are known in *Brassica oleracea*. Here dominance is assumed.

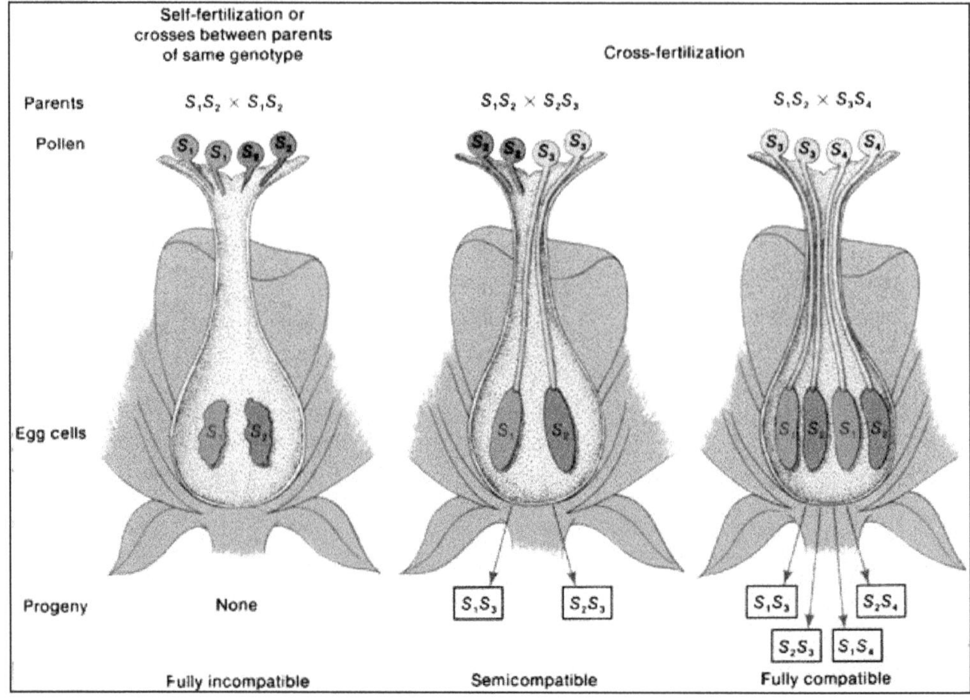

Figure 8: The Compatible Reactions in Gametophytic System of Self Incompatability.

The incompatibility reaction is determined by the genotype of the plant on which pollen grain is produced and not by the genotype of the pollen. This system is more complicated. The allele may exhibit dominance, co-dominance or competition. This system was first reported by Hugues and Babcock in 1950 in *Crepis foetida* and by Gerstal in *Parthenium argentatum*. The sporophytic system is found in radish, Brassicas and spinach.

Lewis has summarized the characteristics of sporophytic system as follows.

1. There are frequent reciprocal differences.
2. Incompatibility can occur with female parent
3. A family can consist of three incompatibility groups.
4. Homozygotes are a normal part of the system
5. An incompatibility group may contain two genotypes.

The letters within parenthesis, *e.g.* (S_1) and (S_2) denote the incompatibility reactions of pollen grains and styles. A complete dominance is assumed.

MECHANISM OF SELF-INCOMPATIBILITY

This is quite complex and is poorly understood. The various phenomena observed in Self-incompatibility are grouped in to three categories.

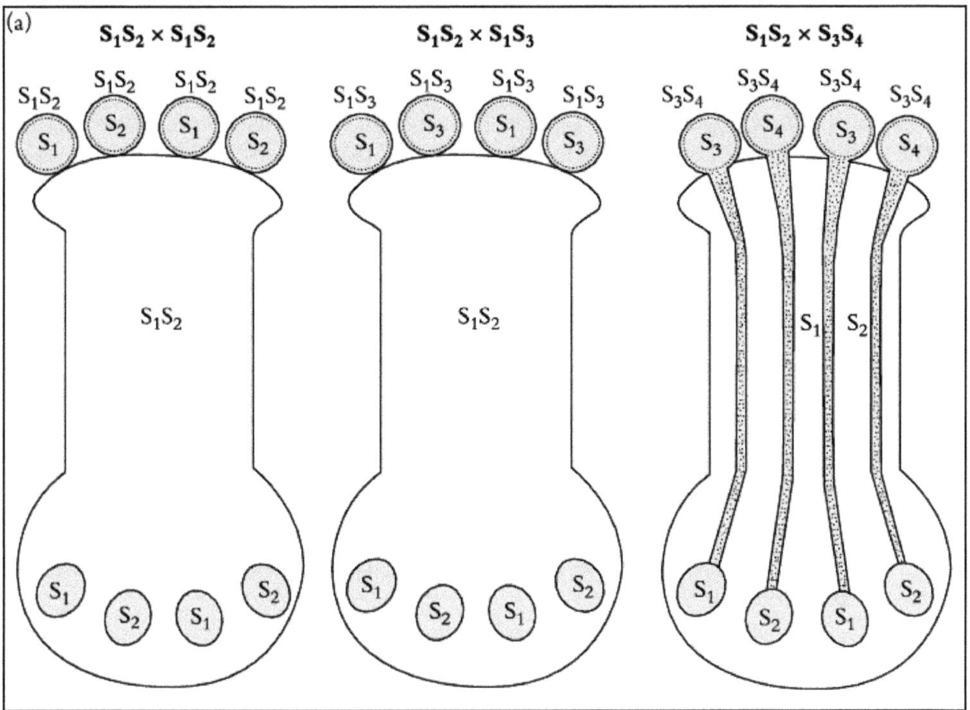

Figure 9: Sporophytic System of Self-incompatibility.

The incompatibility reaction of pollen grains is controlled by the genotype of plant (sporophyte) on which they are produced, while that of style is governed by its own genotype. For simplicity, it is assumed that the incompatibility alleles (*S* alleles) show complete dominance in the manner $S_1>S_2>S_3>S_4$...*etc.*

a) Pollen - Stigma Interaction

This occurs just after the pollen grains reach the stigma and generally prevents pollen from germination. Previously it was thought that binucleate condition of pollen in gamatophytic system and trinucleate condition in sporophytic system was the reason for self incompatibility. But later on it was observed that they are not the reason for SI. Under homomorphic system of incompatibility there are differences in the stigmatic surface which prevents pollen germination. In gametophytic system the stigma surface is plumose having elongated receptive cells which are commonly known as wet stigma. The pollen grain germinates on reaching the stigma and incompatibility reaction occurs at a later stage.

In the sporophytic system the stigma is papillate and dry and covered with hydrated layer of protein known as pellicle. This pellicle is involved in incompatibility reaction. Within few minutes of reaching the stigmatic surface the pollen releases exine exudates which are either protein or glycerol-protein. This reacts with pellicel and induces callose formation which further prevents the growth of pollen tube.

b) Pollen Tube - Style Interaction

Pollen grains germinate and Pollen tube penetrates the stigmatic surface. But in incompatible combinations the growth of pollen tube is retarded with in the style as in *Petunia, Lycopersicon, Lilium*. The protein and poly saccharine synthesis in the pollen tube stops resulting in bursting up of pollen tube and leading to death of nuclei.

c) Pollen Tube - Ovule Interaction

In *Theobroma cacao* pollen tube reaches the ovule and fertilization occurs but the embryo degenerates later due to some biochemical reaction.

Relevance of Self-incompatibility in Plant Breeding

Self-incompatibility may be used for hybrid seed production.

a) By planting two self-incompatible but cross compatible varieties alternatively seeds obtained from both the lines are hybrids.

b) Alternatively by planting a self-incompatible variety along with self-compatible variety, the seeds obtained from self-incompatible line will be a hybrid.

Hybrid seed production was made in Brassicas, Clover, *Trifolium*, Solanaceous and Asteraceae crops.

But there are certain difficulties in this.

a) Production and maintenance of inbred line by hand pollination is tedious and costly.

b) Continuous selfing leads to break down of self-incompatibility and self-fertile lines will appear.

c) Environmental factors such as high temperature and high humidity reduce self-incompatibility.

d) Bees often prefer to stay with in particular parental line which in turn increases the proportion of selfed seed.

Elimination of Self-incompatibility

1) In a single gene gametophytic system by doubling the chromosome number we can eliminate self-incompatibility.
2) By induced mutagenesis to produce self fertile lines.
3) By transferring self compatible alleles from related species through back cross breeding.

Overcoming Self-incompatibility

☆ **By bud pollination:** Application of matured pollen to immature stigma. *e.g. Brassica, Nicotiana, etc.*

- **By surgical technique:** Removal of the stigmatic surface (*e.g.* Brassicas) or removal of style (*e.g. Petunias*).
- **End of season pollination:** In some cases self-incompatibility is reduced towards the end of flowering period. Pollination at that time may be successful (*e.g. Petunias, Nicotiana, etc.)*
- **Use of high temperature:** Exposure of pistil to 60°c will induce pseudo fertility.

  e.g. Malus, Prunus, Trifolium, etc.
- **Irradiation:** Physical agents
- **Grafting:** Grafting of a branch to another branch.
- **Double pollination:** Application of a mixture of incompatible and compatible pollen grains.

5

Male Sterility

Male sterility is the inability of a plant to produce functional pollen grains, while female gametes function normally. It occurs in nature sporadically perhaps due to mutation.

Morphological Features of Male Sterility

The male sterility may be due to mutation, chromosomal aberrations, cytoplasmic factors or interaction of cytoplasmic and genetic factors. Because of any of the above reasons the following morphological changes may occur in male sterile plants.

1. Viable pollen grains are not formed. The sterile pollen grains will be transparent and rarely takes up stain faintly.
2. Non dehiscence of anthers, even though viable pollens are enclosed within. This may be due to hard outer layer which restrict the release of pollen grains.
3. Androecium may abort before the pollen grains are formed.
4. Androecium may be malformed, thus there is no possibility of pollen grain formation.

Male Sterility may exist as

- ☆ Pollen Sterility- in which non-functional pollen is formed.
- ☆ Staminal Sterility- in which stamens are non-functional/even absent.
- ☆ Functional Sterility – in which viable pollen is trapped in indehiscent anthers.

Among these pollen sterility is of most common type. Thus the flowers cannot be self pollinated, but they can be cross pollinated. This attribute makes the male sterile system useful to the plant breeder especially for the production of hybrid seed.

Classification of Male Sterility Systems

1. Genetic Male Sterility
2. Cytoplasmic Male Sterility
3. Cytoplasmic Genetic Male Sterility
4. Transgenic Male Sterility
5. Chemically Induced Male Sterility

A line or ms line: This term represents a male sterile line belonging to any one of the above categories. The A line is always used as a female parent in hybrid seed production.

B line or maintainer line: This line is used to maintain the sterility of A line. The B line is isogenic line which is identical for all traits except for fertility status.

R line and restoration of fertility: It is otherwise known as Restorer line which restores fertility in the A line. The crossing between A x R lines results in F_1 fertile hybrid seeds which is of commercial value.

1. Genetic Male Sterility (GMS)

Failure of pollen production is due to one or more nuclear genes. It is ordinarily governed by a single recessive gene '*ms*' (*i.e.* recessive and monogenic), but dominant genes governing male sterility are also known in safflower. GMS may originate through spontaneous or through induced mutations. It is of wide occurrence in flowering plants and as many as 60 genes for male sterility in Maize, 55 in tomato, 10 in cotton and 60 in rice are known.

The *msms* genotype is male sterile. The dominant allele '*Ms*' results in development of anthers and pollens. Therefore *Msms or MsMs* are male fertile.

Inheritance of Male Sterility

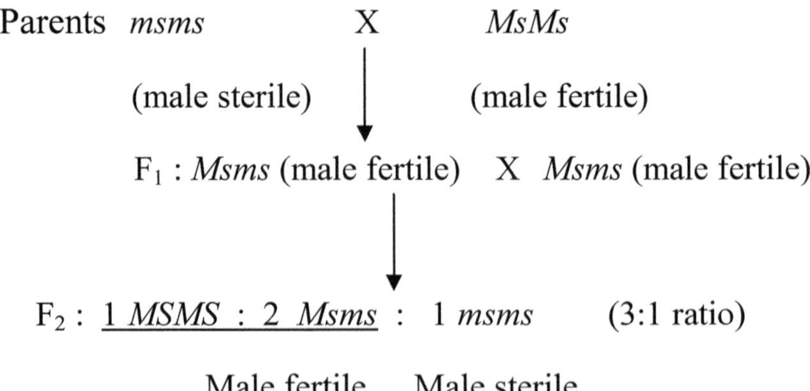

Maintenance of Male Sterile line (A line)

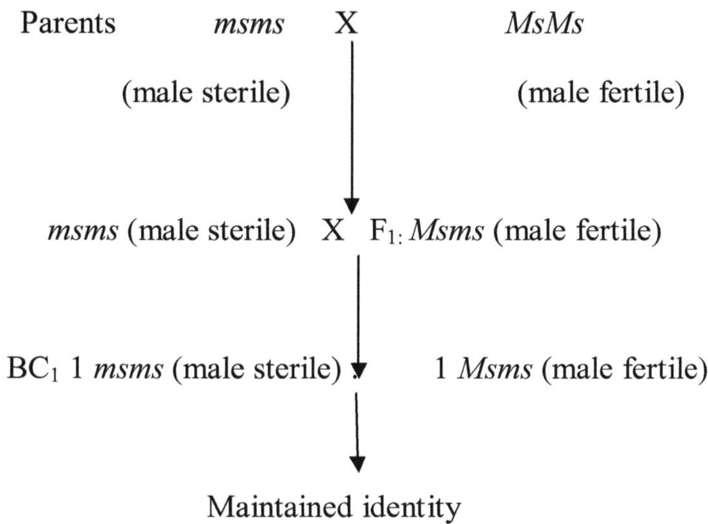

Genetic Male Sterility is subdivided into two Groad Groups

a. Environment insensitive genetic male sterility - *'ms'* gene expression is much less affected by the environment.
b. Environment sensitive genetic male sterility - *'ms'* gene expression is affected by the environment. This is further divided into two groups.

Temperature Sensitive Genetic Male Sterility (TGMS)

In this type of GMS, complete male sterility is produced by the 'ms' gene at higher temperatures (*e.g.* In rice, TGMS line Pei-Ai645, 23.3°C or higher temperature), but at temperatures below this critical point, it exhibits normal fertility.

Photoperiod Sensitive Genetic Male Sterility (PGMS)

In this type of GMS, expression of *'ms'* gene is drastically affected by the prevailing photoperiod provided the temperature is within a critical range. *e.g.* in rice 23-29°C, within this temperature range complete sterility is obtained in plants grown under long day conditions (day length more than 13hr 45 min.), but under short day conditions almost normal fertility is obtained.

Utilization in Plant Breeding

A line is male sterile line, B line is maintainer line

In F_1 generation, roughing off the male fertile plants before anthesis is based on the morphological marker genes closely linked to male sterility locus. Therefore the hybrid seed production becomes costly. Use of TGMS or PGMS eliminates this problem. The male sterile lines are maintained by growing them in a locality where critical temperature and photoperiods exhibits complete fertility. The selfed seeds

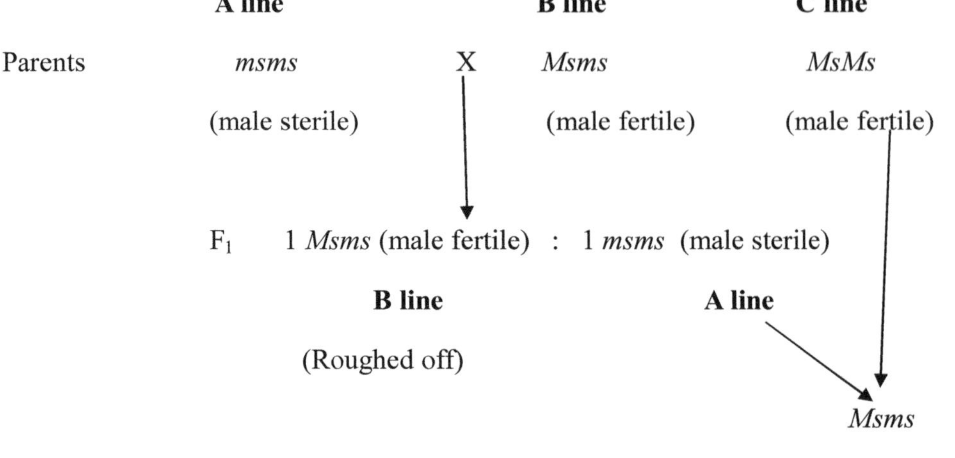

from such lines are then grown in a locality for hybrid seed production where critical temperature and photoperiods exhibits complete sterility. TGMS and PGMS are called as two line breeding methods.

Hybrid Development

e.g., Redgram: T 21 (ms A line) x ICPL 87109 (R line) - CoRH 1

Difficulties in Use of GMS

1. Maintenance of GMS requires skilled labour to identify fertile and sterile line. Labelling is time consuming and costly.
2. In hybrid seed production plot identification of fertile line and removing them is costly.
3. Use of double the seed rate of GMS line is costly.
4. In crops like castor high temperature leads to break down of male sterility.

2. Cytoplasmic Male Sterility (CMS)

The male sterility governed solely by cytoplasmic factors without any dependence on nuclear genes is termed as cytoplasmic male sterility. Since the offspring's inherit the cytoplasm of only the female parent so the cross of cytoplasmic male sterile plants leads to completely male sterile progeny.

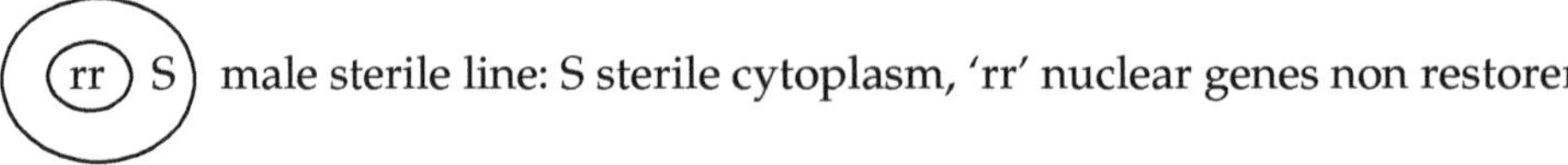

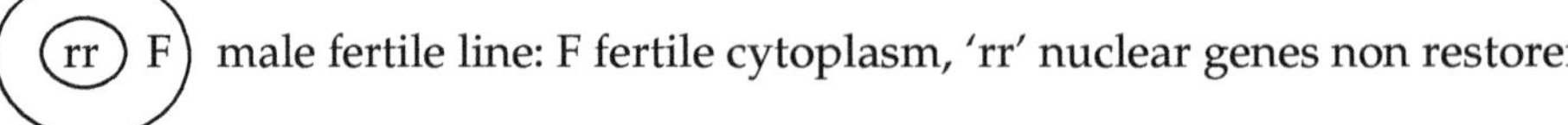

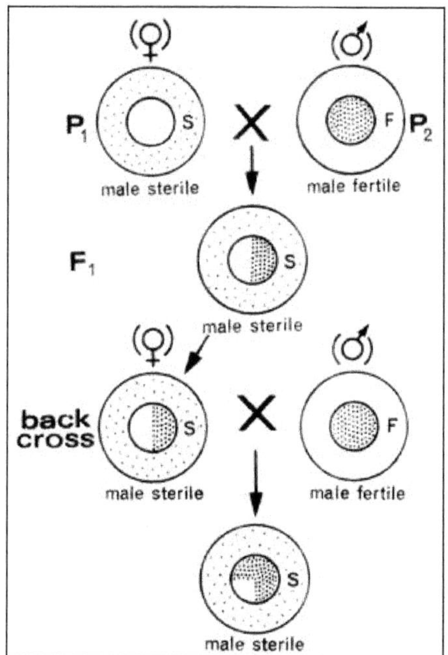

Figure 10: Cytoplasmic Male Sterility System.

Since mother contributes the cytoplasm to the offspring, the sterility is transferred to the F1

Utilization in Plant Breeding

Since there are no R lines available, this type of sterility is useful only in crops where seed is not the nd product. For example in onion and many ornamental plants the hybrids developed exhibit maximum hybrid vigour with respect to longer vegetative duration and larger flower size and larger bulb size. Cytoplasmic male sterility has successfully been exploited in maize for producing double cross hybrids.

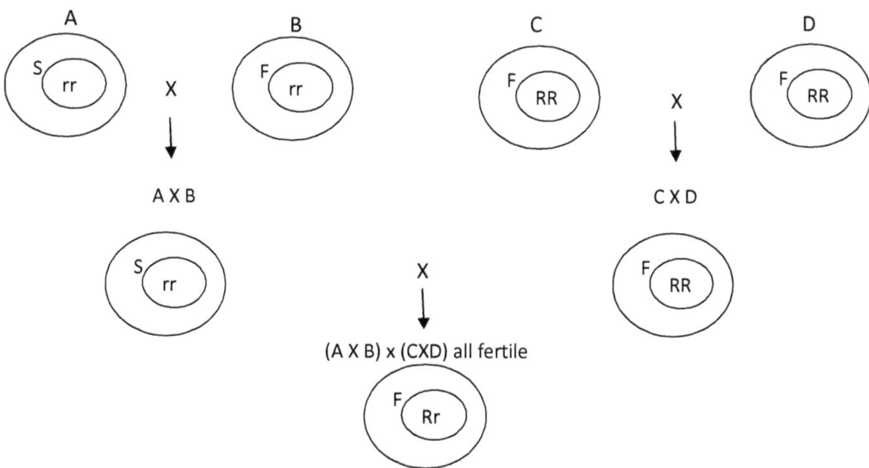

3. Cytoplasmic Genetic Male Sterility (CGMS)

This sterility results from the interaction among the cytoplasmic and nuclear genes. This is a case of cytoplasmic male sterility where a nuclear gene for restoring fertility in the male sterile line is known. The fertility restorer gene *R*, is dominant and found in certain strains of the species or may be transferred from a related species. The cases of cytoplasmic male sterility would be included in the cytoplasmic genetic system as and when restorer genes for them would be discovered. It is being widely used for production of hybrid seed both in self and cross-pollinated crops *e.g.* sorghum, sunflower, pearlmillet, maize, rice and oilseed rape.

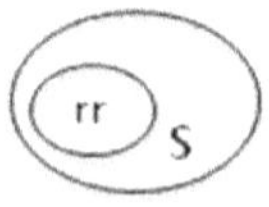

male sterile line: S sterile cytoplasm, 'rr' nonrestorer nuclear genes

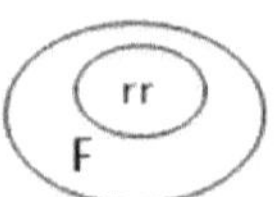

male fertile line: F fertile cytoplasm, 'rr' nonrestorer nuclear genes.

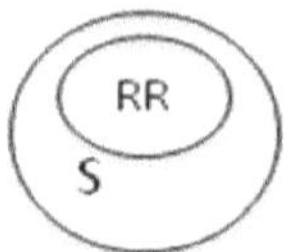

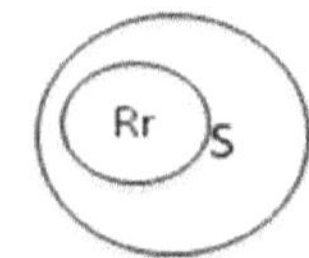

Both are male fertile: S sterile cytoplasm, RR, Rr restorer nuclear genes.

Utilization in Plant Breeding

Generally the use of CGMS for commercial hybrid seed production involves the use of three breeding lines namely, A, B and R line.

A scheme for maintenance of parental lines and commercial hybrid seed production using CGMS is as follows.

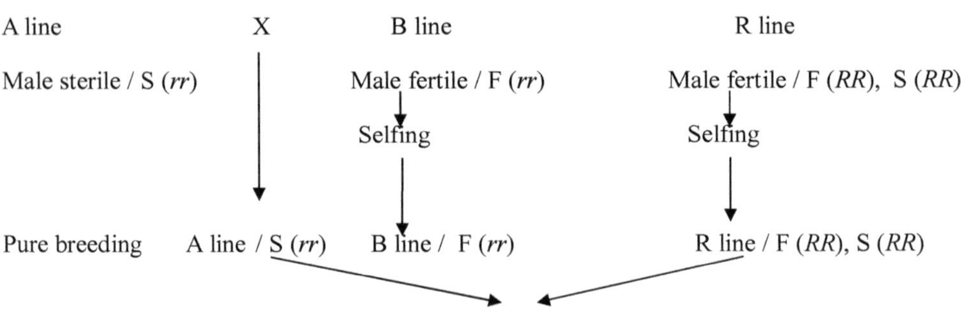

Limitations of CGMS Lines

1. Fertility restoration is a problem. *e.g.* Rice.

2. Seed set will be low in crops like Rice where special techniques are to be adopted to increase seed set.
3. Break down of male sterility at higher temperature.
4. In crops like wheat having a polyploidy series it is difficult to develop effective R line.
5. Undesirable effect of cytoplasm.

   e.g. Texas cytoplasm in maize became susceptible to *Helminthasporium*.

   In bajra Tift 23 A cytoplasm became susceptible to downy mildew.
6. Modifier genes may reduce effectiveness of cytoplasmic male sterility.

4. Transgenic Genetic Male Sterility

Recombinant DNA technology has been successfully used to produce male sterile lines. A gene introduced into the genome of an organism by this technology is called transgene. Many transgenes have been shown to produce genetic male sterility, which is dominant to fertility. However, it is essential to develop effective restoration system, and then these male sterile lines can be used in hybrid seed production. An effective engineered male sterility system is *Barnase* - *Barstar* system in tobacco and *Brassica napus*. Transgenic plants of these crops expressing *barnase* gene were completely male sterile. *Barstar* gene inhibits the activity of *barnase* gene hence restores fertility. Therefore transgenic plants expressing both *barnase* and *barstar* are fully male fertile.

5. Chemically Induced Male Sterility

This type of male sterility is induced by application of certain chemicals. These chemicals are known as male gametocides, male sterilants, pollencides, androcides, pollen suppressants or more commonly known as chemical hybridizing agents (CHA's) which are applied to induce transitory male sterility. The effect these chemicals may range from aberrant microspores meiosis to formation of defective pollen grains. In general CHAs are applied as foliar spray prior to flowering that inhibits the formation of viable pollen but does not affect the seed development. This male sterility is limited to the generation in which the CHA is applied. This approach makes use of two line systems; any parent can be used as female, there is no need of CMS maintainer line and fertility restorer gene. CHA's have been

Table 9: A List of Chemicals Used to Produce Male Sterility

Chemical	*Crop*	*Chemical*	*Crop*
Ethrel	Rice, sugarbeet, wheat	Naphthalene acetic acid (NAA)	Cucurbits
FW 450 (Sodium α,β 2,3, dichloroisobutyrate)	Cotton, groundnut, sugarbeet, tomato	Sodium methyl arsenate	Rice
Gibberellic acid (GA_3)	Lettuce, maize, onion, rice, sunflower	Zinc methyl arsenate	Rice
Maleic hydrazide (MH)	Cucurbits, onion, tomato, wheat	2,4, -D	Rice

tried on cotton, wheat, sorghum, corn *etc.* with varying degree of success. Several commercial rice hybrids have been developed in China. The major problem has been the failure to obtain complete male sterility due to the genotypic response of the crop plants, environmental effects or the differential effects of the chemical itself.

Table 10: List of some Crop Plants in which Male Sterility is Found

Plant Family	*Name of the Plant*		*Type of Male Sterility Found*
	English	*Scientific*	
Gramineae	Barley	*Hordeum vulgare*	GMS
	Wheat	*Triticum aestivum*	GMS, CGMS
	Maize	*Zea mays*	GMS, CGMS
	Sorghum	*Sorghum bicolour*	GMS, CGMS
	Pearlmillet	*Pennisetum americanum*	CGMS
	Rice	*Oryza sativa*	GMS, CGMS
	Sugarcane	*Saccharum offiicinarum*	CGMS
Leguminosae	Lucerne	*Medicago sativa*	GMS
Compositae	Sunflower	*Helianthus anl}uus*	GMS, CGMS
Cucurbitaceae	Watermelon	*Citrullus vulgaris*	GMS
	Muskmelon	*Cucurbita moschata*	GMS
	Pumpkin	*Cucurbita maxima*	GMS
Malvaceae	Cotton	*Gossypium hirsutum*	GMS, CGMS
Solanaceae	Tomato	*Lycopersicon esculentum*	GMS, CGMS
	Tobacco	*Nicotiana tabacum*	GMS, CGMS
Liliaceae	Onion	*Allium cepa*	CMS, CGMS
Chemopodiaceae	Sugarbeet	*Beta vulgaris*	GMS, CGMS
Linaceae	Linseed	*Linum usitatissimum*	CGMS
Umbelliferae	Carrot	*Daucus carrota*	CGMS

LINE BREEDING

The process of using different lines (genotype) and producing hybrid is known as line breeding. This terminology is used in production of rice hybrids.

The different kinds of line breeding are.

a) One line method
b) Two line method
c) Three line method

a) One Line Method of Rice Breeding

Rice hybrids can be developed and propagated through the following concepts.

☆ Vegetative propagation. This can be done by ratooning followed by stubble planting.

- Micropropagation employing tissue culture technique.
- Anther culture hybrids. The anthers of F_1 hybrid can be cultured and plant lets developed.
- Apomictic lines.

b) Two Line Method of Rice Breeding

Two line hybrids can be evolved through application of gametocides and use of environmentally induced genetic male sterility.

To the selected female parent pollen suppressors can be sprayed at the time of flowering so that it will arrest the production of pollen and thus temporary male sterility is induced. The best combiner is used as a male parent and hybrid is produced.

The EGMS system is used successfully in china. Both TGMS and PGMS lines were identified. In this system male sterility is mainly controlled by one or two pairs of recessive nuclear genes and has no relation to cytoplasm. In this system only two lines *viz.* male sterile and Restorer lines are used. Maintainer line is not needed because by growing the male sterile line in suitable atmosphere the sterility is maintained. In this method there is no negative effect due to sterile cytoplasm.

c) Three Line Method or CGMS System

This system nowadays known as CGMS system involving three lines *viz.* a) Cytoplasmic Genic Male Sterile line. b) Maintainer or B line and c) Restorer line.

6

Domestication and Plant Genetic Resources (PGR)

The process of bringing wild species under human management.

Domestication is an evolutionary process operating under the influence of human activities. Being evolutionary, obviously it is relatively a slow process and exhibits gradual progression from the wild state to a state of incipient domestication. Diverse forms that differ more and more from their progenitors develop.

For prominent characters, the evolutionary changes that take place during plant domestication are listed in Table 11 (Hawkes, lecture delivered at the NBPGR, 1978).

Table 11: Evolutionary Changes during Plant Domestication

Sl.No.	*Character*	*Wild Forms*	*Domesticated Forms*
1.	Viability in competition with other species	Good	Poor
2.	Food reserves (size of fruit, *etc.*)	Medium, not very succulent	Large, succulent
3.	Variability of storage organ used by man (size, shape, colour)	Small	Great
4.	Physiological adaptation	Medium to narrow range	Wide range
5.	Dispersal mechanisms such as		
	(a) Rachis or rachilla of cereals	Brittle	Non-brittle
	(b) Stolons in potato	Long	Short
	(c) Explosive dehiscence (legume-pods *etc.*)	Present	Absent
	(d) Pores for seed dispersal (*Papaver somniferum*)	Present	Absent
	(e) Cereal awns	Present	Absent or reduced

Contd...

Table 11–*Contd...*

Sl.No.	Character	Wild Forms	Domesticated Forms
6.	Protective devices, such as		
	(a) Thorns or spines	Present	Absent
	(b) Bitter or poisonous flavour	Present	Absent
	(c) Dense indumentum of hairs	Present	Absent
7.	Sexual reproduction (as in potato, sweet potato, *etc.*)	Present	Absent or reduced
8.	Habit	Perennial	Annual
9.	Seed germination uniformity	Not synchronous	Synchronous and uniform
10.	Breeding mechanism	Outbreeding	Inbreeding

Germplasm

"Choice of germplasm is a critical decision in a breeding program that requires considerable thought; it will determine maximum potential improvement that can be attained via breeding; the breeding system will determine how much of that maximum potential can be realized ." The germplasm collection is a collection of large number of genotypes of a crop species and its wild relatives. **In other words it is the sum total of hereditary or genes present in a species.** Germplasm collections are also known as gene banks or gene pool or world collection.

The germplasm consists of the following 7 types of materials:

1. Land Races

Primitive cultivars, evolved over centuries through both natural and artificial selection, but without a systematic and sustained plant breeding efforts.

- ☆ High level of genetic diversity – diseases, pests.
- ☆ Broad genetic base
- ☆ Less uniform
- ☆ Low yielders.

2. Obsolete Cultivars

Improved varieties of recent past and these were developed by systematic breeding efforts.

- ☆ Replaced by new varieties.
- ☆ Wheat varieties K 68, K65, pb591 – Traditional Tall before Mexican wheat attractive grain color and good chapatti making.

3. Modern Cultivars

- ☆ Currently cultivated high yielding varieties.
- ☆ High yield potential, uniformity, parents in breeding program
- ☆ Narrow genetic base
- ☆ Low adaptability.

4. Advanced Breeding Lines

Pre-released plants developed by plant breeders. Not yet ready for release.

5. Wild Forms

Wild species from which crop species were directly derived.

- ☆ High degree of resistance
- ☆ Easily cross with the concerned crop species.

6. Wild Relatives

Includes all other species, which are related to the crop sps by descent during their evolution.

- ☆ Hybrid sterility problems in crossing
- ☆ Hybrid in viability
- ☆ Undesirable genes with desirable alleles.

Ex: Rice: *O. rufipogon,* Sorghum: *S. halepense,* Peanut: *A. duranensis etc.*

Table 12: The Number of Wild Species of Agri-horticultural Importance (Arora and Nayar, 1984)

Crop Groups	*Number of Wild Species*
Cereals and millets	51
Legumes	31
Fruits	109
Vegetables	54
Oilseeds	12
Fibre plants	24
Spices and condiments	27
Others	26

7. Mutants

Mutant (an organism or cell showing a mutant phenotype due to the mutant allele of a gene) gene pools are valuable genetic resources. *e.g., Dee-Geo-Woo-Gen* in rice and *Norin 10* in wheat. From 1930 to 2014 about 3200 mutant varieties were released (75 per cent filed crops and 25 per cent varieties in ornamental crops).

Classification of Gene Pool

It consists of all the genes and their alleles present in all such individuals, which hybridize or can hybridize with each other.

1. **Area of collection:** a) Indigenous; b) Exotic collections
2. **Domestication:** a) Cultivated; b) Wild gene pool
3. **Cross ability in Breeding Program:** (Harlan and De Wet, 1971)

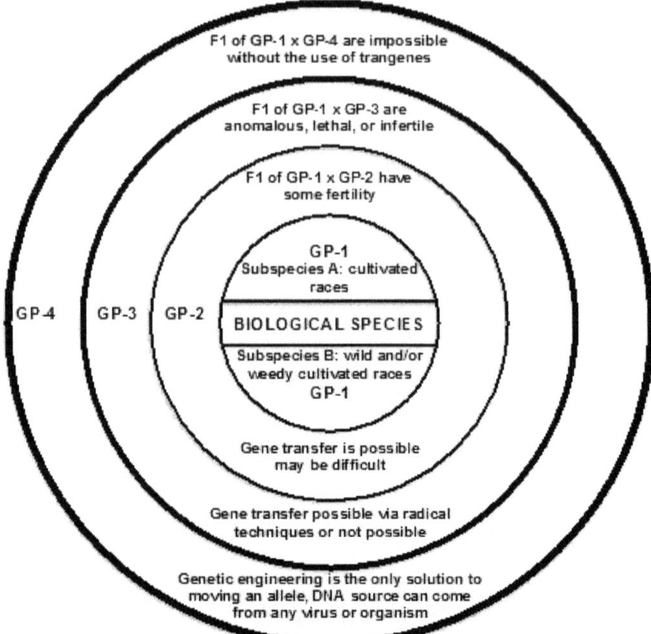

Figure 11: Genepool Concept of Harlan and De Wet, 1971.

1. Primary gene pool (GP-1): Inter-mating is easy and production of fertile hybrids. Hybridization between the same species or closely related.
2. Secondary gene pool (GP-2): Partial fertility on crossing with GP1 plants related species.
3. Tertiary gene pool (GP-3): Sterile hybrids on crossing with primary gene pool and needs special techniques, to obtain the fertile F1s.
4. Gene pool (GP-4): Sources of genes across general species and kingdom. It is also called a gene ocean. *e.g: Bt* genes from *Bacillus thuringiensis*.

Threat to Genetic Diversity

Genetic vulnerability: High input & monoculture over large areas

Genetic wipe out: Rapid & wholesale destruction of potential species.

Genetic erosion: The gradual loss of variability in the cultivated forms and in their wild relatives is referred to as genetic erosion.

Factors Responsible for Genetic Erosion

a) Replacement of land races with improved cultivars
b) Modernization of agriculture – eliminates wild and weedy former.
c) Extension of farming into wild habits- shifting cultivation
d) Grazing into wild habitats.

Table 13: Genetic Composition, Productivity Level and Potential Value in Breeding of different Gene Sources (Chang, 1985)

Group	*Diversity within Group*	*Homogeneity within a Strain or Population*	*Agronomic or Commercial Value*	*Genetic Potential in Breeding*
Modern elite cultivars	Low to moderate	Very high	Very high	Moderately high
Principal commercial types	Moderately low to moderate	Moderate to high	Moderately high	Moderate
Minor varieties	Moderately high to high	Moderately low to moderate	Moderate	Moderately high
Speciality types	Moderately low to moderate	Moderately high	Moderately low	High
Obsolete types	Moderate to high	Moderately high to high	Moderately low to moderate	Moderately low
Breeding stocks	Moderately low to moderately high	Moderate for lines; low for bulks	Most variable	Moderate to high
Mutants	Moderately low to moderate	Moderately high to high	Mostly low; few moderately high to high	Mostly low
Primitive types/land races	Moderately high to high	Low to moderate	Moderately low	Moderately high to high
Weed races	Moderately high to high	Low to moderately low	Low	Moderate to moderately high
Wild species	Moderately low to moderate	Low to moderately low	Very low	Moderate to high

e) Growth of cities/urbanization.
f) Change in agricultural situations
g) Environmental changes
h) Outbreak of pest and diseases
i) Sivil strife *etc.*

With the modernization of agriculture, large tracts of land have been put under pure line varieties of self-pollinated crops and hybrid varieties of cross-pollinated species. This has led to a gradual disappearance of local or land varieties ('desi' varieties) and open- pollinated varieties (both reservoirs of considerable variability). Cultivation and grazing are gradually destroying many wild species and their breeding grounds. Wild relatives of crops may be eliminated by introduced species of weedy nature or even by the cultivated forms derived from them.

This variability arose in nature over an extremely long period of time and, if lost, would not be reproduced during a short period. Most of the countries are greatly concerned about genetic erosion. The establishment of IBPGR to coordinate germplasm conservation activities throughout the world reflects this concern. Germplasm collections are being made and maintained to conserve as many genotype as possible. The germplasm collections contain land varieties, various wild forms, primitive races, exotic collections and highly evolved varieties.

Need for Germplasm Bank

a) The modernisation of agriculture and evolution of high yielding varieties and hybrids led to the replacement of the land races.
   - ✰ For examples after the introduction of IR 20 rice for samba season all the local varieties like Karthigai Samba, Toppi Samba, Rubber Samba, Thiruchengodu Samba and Athur Samba went out of cultivation. Along with them the beneficial genes also vanished. So, to prevent the genetic erosion it is necessary to maintain germplasm.

b) Nature has provided enormous variability for the use of mankind. We should not destroy them and preserve them for the use of future mankind.

   * Extinction: Permanent loss of a crop species.

Germplasm Activities

See *Figure 12.*

1) Exploration and Collection

Expolration: Collection trips tapping genetic diversity from various sources.

Collection: assembling of tapped genotypes at one place.

Process

i) Sources of collection:
   - ✰ Centres of diversity

Figure 12: Germplasm Activities.

- ✰ Gene bank
- ✰ Gene sanctuaries
- ✰ Seed companies
- ✰ Farmers fields

ii) Priority of collection: Endangered areas, endangered species.

Table 14: Crop Collecting Priorities in National Context (Arora, 1988)

High	*Moderate*	*Low*
Wheat, rice	Maize, barley	–
Sorghum, finger millet, pearl millet	Prosomillet and others	Amaranth, buckwheat
Chickpea, pigeonpea, *Vigna* (Asiatic)	Lentil	Soybean, winged bean, rice bean
Groundnut, sesame, Brassicae	Safflower	Niger
Cotton	Jute	Kenaf/mesta
Citrus, banana	Mango, jackfruit, egg plant, okra, cucurbits, tuber crops	–
Coconut, sugarcane	Tea, arecanut, turmeric, medicinal plants	Ginger

iii) Agencies of collection:

- ✰ SAU
- ✰ ICAR
- ✰ IPGRI

iv) Method of collection: Expeditions, personal visit to gene bank, correspondence, exchange of material.

v) Method of sampling:

- Random sampling – biotic tolerant stresses
- Biased sampling – distinct morphological characters.

vi) Sample size: 50 –100 individuals, 50 seeds/plant. It should represent 95 per cent of total diversity.

Merits

- Tapping crop genetic diversity
- New material
- Prevents extinction

Demerits

- Entry of new disease.
- Tedious, time taking and expensive
- It poses various hardships to the collectors

Some of the Important Germplasm Collections are listed Below

1. Institute of Plant Industry, Leningard. It has 1,60,000 entries of crop plants.
2. Royal Botanic Gardens, Kew, England, It has over 45,000 entries.
3. IRRI, Philippines, is maintaining 42,000 rice strains and varieties. More than 15,000 entries are maintained at NRRI, Cuttack.
4. World collections of some of the crops are maintained at the following places.
    i) Sugarcane: Canal Point, Florida, U.S.A. and Sugarcane Breeding Institute, Coimbatore (2,800 entries).
    ii) Groundnut: Senegal (Africa), ICRISAT, Hyderabad, BARC, Mumbai, DGR, Junagadh.
    iii) Potato: Cambridge, U.K. and Wisconsin, U.S.A.
    iv) Annual New World Cottons: Near Tashkent, U.S.S.R.
    v) Coffee: Ethiopis (Africa).
    vi Sweet Potatoes: New Zealand
5. Bellsville, U.S.A., maintains germplasm collections of small grain crops.
6. The National Bureau of Plant Genetic Resources, New Delhi, is maintaining large collections of Sorghum, Pennisetum, wheat, barley, oats, rice, Maize and other agricultural and horticultural crops.
    - For example, groundnut collection is maintained at Junagarh, Cotton at Nagpur, Potato at Simla, Tobacco at Rajahmundhry, tuber crops (other than potato) at Trivandrum *etc.*

- ☆ The Cotton collection maintained at Central Institute for Cotton Research (CICR, Nagpur) are as follows; *Gossypium hirsutum*- 4,100 entries; *G. barbadense*- 300 entries; *G. arboreum*- 1755 entries; *G. herbaceum*- 393 entries (1991).

7. The various International Institutes are building up and maintaining collections of many species.

2. Conservation

Seeds of most species lose viability quickly. Consequently, germplasm collections have to be grown every few years. (1) Growing, harvesting and storing large collections is a costly affair requiring much time, labour, land and money. (2) There is also risk of errors in labeling. (3) The genotypic constitution of entries may also change, particularly when they are grown in environments considerably different from that to which they are adapted. This is particularly true in case of cross-pollinated species and for local varieties of self- pollinated species. These difficulties may be considerably reduced by cold storage of seeds. Seeds of most of the plant species can be stored for 10 years or more at low temperatures and low humidity. Thus the entries could be grown every 10 years or so instead of every one or two years. Cold storage facilities are being utilized at Fort Collins, U.S.A and at IRRI, Phillippines, NBPGR has developed cold storage facilities for germplasm maintenance this is known as National Germplasm Repository.

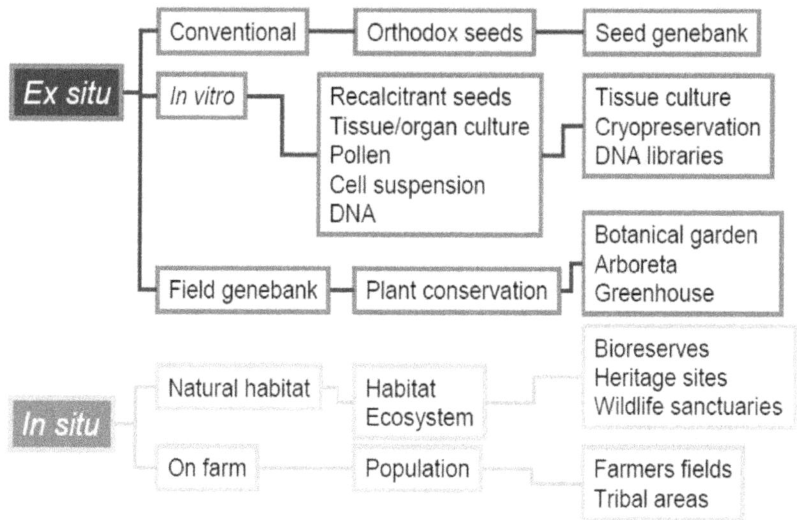

Figure13: Methods of Germplasm Conservations

Types of Conservation

a) Based on Duration of Conservation

i. Short Term Conservation

- ☆ Conservation for 3-5 years at 5-10° C and seed moisture 8-10 per cent.

- ☆ Based on the viability of the seeds the gene pool is to grown once in two years or more than two years. Each line is to be grown with proper spacing and care must be taken to ensure self pollination, so that the genetic architecture is not altered. *For example* in sorghum covering the panicle in boot leaf stage itself ensures selfing.
- ☆ This short term conservation is a costly affair which requires much time, labour, land and cost. Further there is every chance for mixing up of genotypes while large number is handled annually.

ii. Medium Term Conservation

- ☆ Conservation for 10-15 years at 0°C and seed moisture 8 per cent.
- ☆ Routine tests after 5-10 years
- ☆ Comparatively less costly than short term conservation.

iii. Long Term Conservation

- ☆ Conservation for 10-50 years at -18°C to -20°C and seed moisture 5-6 per cent.
- ☆ To overcome this difficulty long term preservation in the cold storage the germplasm can be preserved. Using liquid nitrogen (-196° C) the germplasm can be stored for more than ten years. Complete information about the genotypes can be computerized and this is known as cataloguing and information retrieval system.

b) Based on Methods or Place of Conservation

i. In-situ Conservation: Under natural condition only

The areas of diversity are protected from trespass of human beings by fencing the area so that the plant species are preserved under natural conditions. This is known as *in-situ* conservation.

- ☆ Gene Sanctuaries

  e.g: - Meghalaya for citrus

  - North Eastern Region for *Musa, Oryza, Saccharum.*
- ☆ Biosphere reserves.
- ☆ National parks
- ☆ Botanical gardens *etc,*

  * Costly method, several areas have to be conserved.

  * Protection of genetic diversity of plants from genetic erosion.

ii. Ex-situ Conservation

Preservation of germplasm in gene banks, mass reservoirs, field gene banks, cold storage, in-vitro conservation.

- ☆ Cheaper, easy, entire genetic diversity conserved.

3. Evaluation

1) To identify gene sources
2) Classification of germplasm.
    - ✰ By simple measures of dispersion (Range, standard deviation, SE, CV)
    - ✰ By metroglyph analysis of Anderson (1957)
    - ✰ D^2 statistic of Mahalanobis (1936), *etc.*

4. Documentation

- ✰ Provides information about various activities of plant genetic resistance (compilation, analysis, classification, storage, dissemination of information).

  e.g: 7.3 million germplasm accession – 200 crop species.

5. Distribution

Specific germplasm – supplied on demand.

6. Utilization

Use of germplasm in crop improvement programme

- ✰ As a variety
- ✰ As parent
- ✰ As variant in gene pool.
- ✰ Wild – Transfer of resistance genes.

7

Centres of Diversity/Origin

The cultivation of plants is one of man's oldest occupations and probably began when he selected some plants for his use. One of the old belief regarding to the origin of cultivated plants was that they came to man as a gift from God. By the end of 18th century people started questioning about the origin of cultivated plants. Knowledge of origin is important in order to avoid genetic erosion and also helps in locating wild relatives, related species and new genes.

Definition

Centres of Origin/Centres of diversity is a geographical area where group of organisms, either domesticated or wild, first developed its distinctive properties.

Darwin (1868) considered that the cultivated plants arose by profound modifications in the wild plant.

Alphonse de Candolle (1863) a Swiss botanist first attempted to solve the mystery about evolution of crop plants. In his "Origin of Cultivated Plants" he studied 247 plant species of cultivated plants. He classified the economic plants into six classes;

1. Plants cultivated 4000 years ago.
2. Plants cultivated less than 4000 years.
3. Plants cultivated 2000 to 4000 years.
4. Plants cultivated 2000 years ago.
5. Plants cultivated before the time of Columbus (1492).
6. Plants cultivated after the time of Columbus.

It is ***Nikolai Ivanovich Vavilov* (1925),** who proposed the concept of '**Centres of Origin**'. He proposed the concept based on his studies of a vast collection of plants at Institute of Plant Industry, Leningrad. The concept is that crop plants evolved from wild species in the area showing great diversity and that place is termed as **Primary Centre of Origin**. Later on from the primary centre the crops moved to other places due to the activities of man.There are certain areas where some crops exhibit maximum diversity of forms but this may not be the centre of origin for that particular crop. Such centres are known as **Secondary Centres of Origin**.

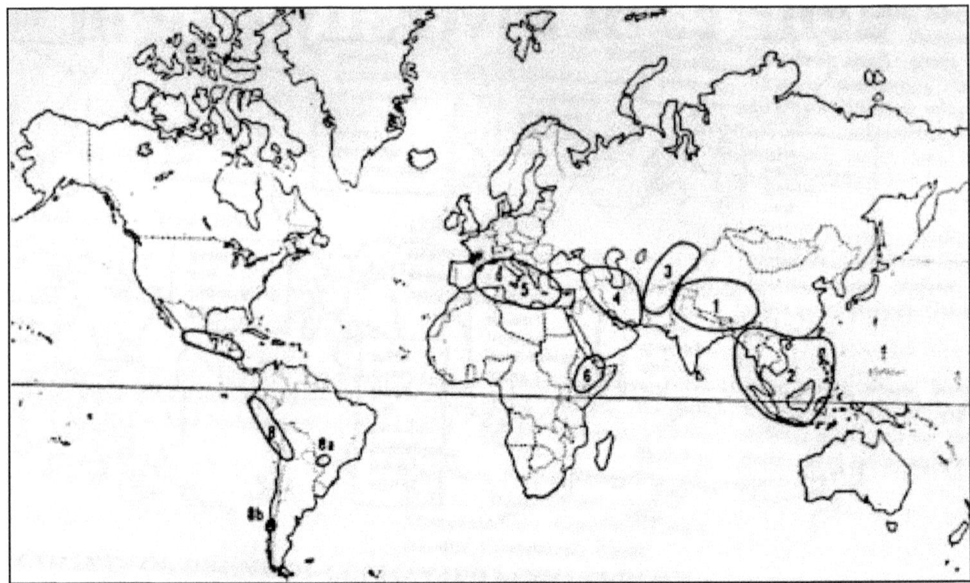

Figure 14: Vavilov's Eight Main Centres of Origin.

Based on his study proposed the eight main centres of origin and they are,

1. The China Centre

It consists of the mountainous regions of central and western China and the neighbouring low lands. It is the largest and oldest independent centre (Table 15).

2. The Hindustan Centre

This includes Burma, Assam, Malaya, Java Borneo, Sumatra and Philippines, but excludes North West India, Punjab and North Western Frontier Provinces (Table 16).

3. The Central Asia Centre

It includes North West India, all of Afghanistan, the Soviet Republics of Tadjikistan and Tian Shan (mountain range in Aksu, Xinjiang Asia). It is also known as *the Afghanistan centre of origin* (Table 17).

Table 15: Important Crops Originated in the China Centre

I.Primary centre of origin are:	
Soybeans	Chinese apple
Radish	Pear
Proso millet	Apricot
Opium	Litchi
Brassica	Walnut
Onion	Cherry
Soyabean	Camphor
Adzuki bean	Hemp
Cabbage	***ii. Secondary centre of origin are:***
Onion	Maize
Chinese yam	Cowpea
Sugar cane	Turnip
Opium poppy	Sesame
A total of 136 endemic plants are listed	

Table 16: Important Crops Originated in the Hindustan Centre

Primary centre of origin are:	
Rice	Cucumber
Redgram	Radish
Chickpea	Brinjal
Cowpea	Noble canes
Blackgram	Cotton (*Gossypium arboreum*)
Greengram	Hemp
Turmeric	Coconut
Mango	Tamarind
Orange	Toro
Citron	Yam
A total of 172 plants are listed	

4. The Asia Minor Centre

This is also known as the *Near East orMiddle East or the Persian Centre* of Origin. It includes the interior of Asia Minor, the whole of Transcaucasia, Iran and Highlands of Turkmenistan (Table 18).

5. The Mediterranean Centre

The crops originated in this centre are: Many valuable cereals and legumes such as showing in Table 19.

Table 17: Important Crops Originated in the Central Asia Centre

i. Primary centre of origin are:	
Wheat	Lentil
Pea	Horsegram
Broad bean	Mungbean
Green gram	Mustard
Sesame	Carrot
Safflower	Apple
Almond	Grape
Cotton (*G.herbaceum*)	Hemp
Onion	***ii. Secondary centre of origin are*:**
Garlic	Rye
A total of 43 plants are listed	

Table 18: Important Crops Originated in the Asia Minor Centre

i. Primary centre of origin are:	
Triticum	Barley (two-row)
Rye	Fig
Alfalfa	Pomegranate
Cabbage	Apple
Oats	Cherry
Clover	***ii. Secondary centre of origin are*:**
Fenugreek	Rape
Lentil	Black Mustard
Lupin	Turnip
A total of 83 plants are listed	

Table 19: Important Crops Originated in the Mediterranean Centre

Primary centre of origin are:	
Durum Wheat	Chikpea
Emmer Wheat	Beets
Barley	Peppermint
Lentil	Broad bean
Pea	Oats
Cabbage	Asparagus
Turnip	Grasspea
Lettuce	Flax
Peppermint	Clover.
A total of 84 plants are listed	

6. The Abyssinian Centre/Ethiopia Centre

It includes Ethiopia and hill country of Eritrea and part of Somaliland (Table 20).

Table 20: Important Crops Originated in the Abyssinian Centre

i. Primary centre of origin are:	
Barley	Cowpea
Sorghum	Sesame
Pearl millet	Castor
Lentil	Flax
Khesari	Bean
Sunflower	Coffee.
Castor	***ii. Secondary centre of origin are:***
Coffee	Broad bean
A total of 38 plants are listed	

7. Central American Centre

This includes South Mexico and Central America (Mexico, Guatemala, Hondurus and Costa Rica). It is also referred to as the *Mexican Centre of Origin* (Table 21).

Table 21: Important Crops Originated in the Central American Centre

Primary centre of origin are:	
Maize	Sweet Potato
Lima bean	Arrowroot
Melons	Cotton (*G.hirsutum*)
Pumpkin	Sisal
Papaya	Pepper
Guava	Cocoa
Cashew	Black berry.
A total of 24 plants are listed	

8. The South American Centre

This centre includes the high mountainous regions of Peru, Bolivia, Ecuador, Colombia, parts of Chile, and Brazil and whole of Peraguay (Table 22).

Later in, **1935,** ***Vavilov*** divided the Hindustan Centre of Origin into two centres, *viz., Indo-Burma* and *Siam-Malaya-Java Centre of Origin*. The South American Centre was divided into three centres, namely, *Peru, Chile* and *Brazil-Peraguay Centres of Origin*. At the same time he introduced a new centre of origin, the *U.S.A. Centre of origin*. Two plant species, Sunflower (*Helianthus annuus*) and Jerusalem Artichoke (*H. tuberosus*) originated in the U.S.A. Centre of Origin.

Table 22: Important Crops Originated in the South American Centre

Primary centre of origin are:	
Potato	Tobacco
Maize	Egyptian cotton (*G.barbadense*)
Lima bean	Tapioca
Peanut	Pumpkin
Common bean	Pepper
Cocoa	Pineapple
Rubber	Cashew
A total of 62 plants are listed	

Thus the centres of origin may be more appropriately called the centres of diversity. The centres of origin may not be the centres of origin of the species concerned, but they are the areas of maximum diversity of the species. Within the large centres of diversity, small areas may exhibit much greater diversity than the centre as a whole. These areas are known as ***Microcentres.***

Law of Homologous Series

This is proposed by N. I. Vavilov. According to this law "the characters found in one species also observed in other related species". Thus diploid, tetraploid and hexaploid wheat show a series of identical characters. So also in case of diploid and tetraploid cotton. Similarly genus *Secale* duplicates the variation found in *Triticum.*

Objections to Vavilov's Theory

According to ***Vavilov,*** whenever a crop plant exhibits maximum diversity, that place is the centre of origin for that crop, but this view is no longer valid in case of maize and tomato. For maize the centre of diversity is Peru but archeological evidence shows Mexico as centre of origin. For tomato, South America is considered to be primary centre of origin but it is Mexico as per archeological evidence.

Secondly, Vavilov stated that primary centre is marked by a high frequency of dominant genes in the centre and recessive genes towards the periphery, but it is not so. *e.g.* wheat, maize, oil palm, *etc.*

Vavilov's claim that centre of origin confined to mountainous regions only. But this is not the case. *For example* maize exhibits maximum diversity in plains.

Many crops have more than one centre of origin (*e.g.*Balsam, Sorghum). In some crops centre of domestication cannot be determined for want of suitable evidence.

Megagene Centres

To counter the objection, **Zhukovsky** student of Vavilov has proposed '**Mega Centre**' theory. He divided the world into 12 regions. Mega gene centres were the places where cultivated plant species exhibit diversity and micro gene centre is the place where wild species occur.

Zhukovsky (1965), a close associate of Vavilov, proposed 12 mega gene centres of crop-plant diversity. The new areas added to Vavilov's eight centres were Australia, whole Africa and Siberia followed by revision of the boundaries to make 12 centres. **Micro Gene Centres** of wild growing species related to our crop plants, where the cultigens first originated, were also demarcated.

Zeven and Zhukovsky (1975) have dealt this elaborately in the *'Dictionary of Cultivated Plants and their Centres of Diversity'*, listing species for different mega gene centres, and the range and extent of the distribution of genetic/varietal/specific diversity, *etc.* In a further revised version of this book, Zeven and de Wet (1982) prefer the term region to centre. These twelve regions have by and large wider coverage and more acceptability.

Centres and Non-centres

Harlan (1971) stated that each crop may have been repeatedly domesticated at different times in different locations or may have been brought into cultivation in several regions simultaneously. We cannot pin point a single centre of origin. Harlan developed the idea of '**Centre'** and **'Non- centre'**. According to him 'centre' means places of agricultural origin and 'non centre' where agriculture has been introduced. Harlan divided the world in to three centres and three non centres.

Table 23: List of Centres and Non-centres (Horlan, 1971)

Centre	*Non-centre*
North Chinese Centre - B1	South-East Asian and South Pacific non-centre - B2
North East Centre - A1	African non-centre - A2
Meso American Centre - C1	South American non-centre - C2

Table 24: The Nuclear Centres and Regions of Diversity

Nuclear Centres	*Regions of Diversity*	*Outlying Minor Centres*
A. Northern China	I. China	1. Japan
	II. India	2. New Guinea
	III. South-East Asia	3. Solomon Islands, Fiji and South Pacific
B. The Near East	IV. Central Asia	4. Northwestern Europe
	V. The Near East	
	VI. The Mediterranean	
	VII. Ethiopia	
	VIII. West Africa	
C. Southern Mexico	IX. Meso-America	5. United States, Canada
		6. The Caribbean
D. Central to Southern Peru	X. Northern Andes	7. Southern Chile
	(Venezuela to Bolivia)	8. Brazil

Harlan also recognised smaller areas/pockets of varietal and/or racial diversity within a Vavilovian Centre, and he termed these as 'Micro centres'. Such small areas, as in Turkey and Africa (Harlan, 1975), contain varietal diversity of several crops in the plains and/or mountains.

Nuclear Centres and Regions of Diversity

Hawkes (1983) suggested that generally agriculture began not once but several times, more or less simultaneously and in different regions of the world. His concept clearly envisaged centres of agricultural origin from which farming spread into one or more regions for which he proposed the name 'Nuclear centres and regions of diversity'. He linked the nuclear centres with the archaeological evidence to provide strong proofs of agricultural origins. According to him, the following nuclear centres and regions of diversity occur.

8

Biometrics in Plant Breeding

Variation

When different plants/lines of a species show different magnitudes of a character, it is called variation. Variation is essential for any improvement in a species. Therefore, the first step in any breeding programme is to create variation if it is not already present in the population to be improved. Variation can be created by hybridization, mutation, polyploidy, domestication, plant introduction, somaclonal variation, genetic engineering *etc.*

Biometry/Biometrics

Is the science that deals with the application of statistical procedures to the study of biological problems.

Biometrical Genetics

It is a branch of genetics, which utilizes various statistical concepts and procedures for genetical studies, is called biometrical genetics or also known as quantitative genetics.

Biometrical Techniques

The various statistical procedures employed in a biometrical genetics are called biometrical techniques.

The two basic requirements of plant breeding are the presence of genetic variation and exploitation of this variation through selection. The selection of plants from a population is almost always based on their phenotype. Phenotype has both heritable and non heritable components. The heritable component is due to the genes present in plants, that is genotype. The non- heritable component consists of the effects of environment. The value of progeny obtained from a selected plant, therefore, would largely depend upon the relative contributions by the heritable and the non-heritable components to its phenotype. Clearly the breeder should be thoroughly familiar with the laws of inheritance and the relative importance of the genotype and the environment in determining the concerned phenotype.

Qualitative and Quantitative Characters

The inheritance of both qualitative and quantitative characters follows the laws of Mendel. But the effects of individual genes in the two cases are totally different in magnitude consequently; the techniques used to study the two types of characters are also different.

Table 25: Major difference between the Quantitative and Qualitative Traits

Features	*Quantitative*	*Qualitative*
Variation	Continuous	Discontinuous
No. of genes	Several (Polygenes)	Few (Oligogenes)
Effect of single gene	Minor or small	Major
Classification	Into different clear cut groups is not possible	Possible, due to discrete or discontinuous variations
Gene action	Additive gene action usually but dominance and epistasis also	No additive (dominance and epistasis)
Environmental effect	Considerable influence	Not much influence
Metric measurements	Possible, like size, weight, duration, *etc.*	Not possible, only counting, ratios, *etc.* possible.
Transgression segregates	Possible, from the crosses between two parents with mean values for a quantitative traits	Not possible
Stability	Less stable, because of sensitive to environment	Stable
Field of genetics	Biometrical or quantitative	Mendalian genetics
Heritability estimates	Low, because of high environmental variations	High
Statistical parameters	Mean, variance, co-variance, *etc.*	Segregation ratios, fractions *etc.*

Role of the Environment in Quantitative Inheritance

Quantitative characters are considerably affected by environment. The main result of this effect is that the relationship between genotype and phenotype is partially or completely hidden *i.e.* the phenotype does not reveal the genotype.

e.g.: Phenotype = Genotype + Environment + (G x E)

$$P = G + E + (G \times E)$$

If environment influence = 0, then phenotype = genotype. However the effect of environment is seldom zero. So, phenotype is the joint action of genotype and environment.

In crop improvement, the breeder selects plants on the basis of their phenotype. The effectiveness of selection depends on the proportion of phenotype due to the genotype. Therefore, it is important to know the extent to which environment influences different quantitative characters. To estimate the effect of environment on a character, large Number of strain/genotypes are grown in a replicated trial and the data is subjected to analysis of variance as per the experimental design used.

The genotype x environmental interaction signifies that the relative performance of various genotype in effected by the environment. For *e.g.*: Performance of genotype 'A' may be superior to the genotype 'B' in one environment but in another environment inferior to that of 'B'. If G x E interaction is absent, genotype 'A' will be superior over genotype 'B' in all the environments.

Biometrical Techniques in Plant Breeding

The biometrical techniques are useful to the plant breeders in the following 4 different ways.

1. Assessment of genetic variability present in the population.
   - ✰ It can be assessed by Range, Variance, Standard Deviation, Coefficient of Variation, Correlation Analysis, Regression Analysis, D^2 Statistics and Metro-glyph Analysis, Path Coefficient Analysis *etc.*
2. In the selection of elite genotypes from mixed populations
   - ✰ Correlation, Path and Discriminate function analysis
3. Selection of parents and breeding procedures
   - ✰ Diallel Cross, Partial Diallel Cross, Line x Tester, Triallel Cross, Quadriallel Cross, Generation Mean Analysis, Biparental Cross and Triple Test Cross analysis.
4. Determining Varietal Adaptation: Stability Analysis Models:
   - i. Finley and Wilkinson Model (1963)
   - ii. Eberhart and Russel Model (1966)
   - iii. Perkins and Jinks Model (1968)
   - iv. Freeman and Parkins Model (1971)
   - v. AMMI Model – Additive Main effects and Multiplicative Interaction.

Components of Genetic Variance

Variation

Variability refers to the presence of differences among the individuals of plant population. Variability results due to differences either in the genetic constitution of the individuals of a population or in the environment in which they are grown. The extent of variability is essential for resistance to biotic and abiotic factors as well as for wide adaptability. The magnitude of variability present in a crop species is of utmost importance as it provides the basis for effective selection. The variation present in the plant population is of three type's *viz.* phenotypic, genotypic and environmental variation.

Phenotypic Variation

The total variation present in a population is called phenotypic variation. This arises due to genotypic and environmental effects. Phenotypic variability is the observable variation present in a character in a population. It includes both genotypic and environmental components of variation. As a result, its magnitude differs under different environmental conditions. Such variation is measured in terms of phenotypic variance.

Genotypic Variation

This variation is due to the genotypic differences among individuals within a population and is more useful to the plant breeder for exploitation in selection or hybridization. It is the heritable portion of total variation or phenotypic variation which is unaltered by environmental conditions. This variation is measured in terms of genotypic variance.

Environmental Variation

It refers to non-heritable variation which is entirely due to environmental effects and varies under different environmental conditions. This uncontrolled variation is measured in terms of error mean variance. The variation in true breeding parental lines and their F_1 is non-heritable.

Phenotypic variance = Genotypic variance + Environment variance.

$$Vp = Vg + Ve$$

Phenotypic variance = $\delta^2 p = \delta^2 g + \delta^2 e$

Fisher (1918) divided the genotypic variance into three components: Additive, Dominance and Epistasis (Table 26). Later Hayman and Mather partitioned the epistatic component into three types of interactions – *viz.*, Additive x Additive; Additive x Dominance and Dominance x Dominance (Table 27).

Table 26: Difference between the Additive, Dominance and Epistasis Variances

Additive Variance	*Dominance Variance or Intra Allelic Interaction*	*Epistasis Variance or Inter Allelic Interaction*
Arises of from difference between two homozygotes for a gene *i.e.* *AA* and *aa*.	It is due to the deviation of heterozygote (*Aa*) phenotype from the average of phenotypic values of the two homozygotes (*AA* and *aa*)	Deviation from the additive scheme as a consequence of inter-allelic interactions or interaction between two or more genes
It is generally represented as 'd'	It is represented by 'h'.	Represented by 'e'.
Genes show lack of dominance *i.e.* intermediate expression	Genes show incomplete, complete or over Dominance	Includes both additive and non-additive components
Associated with homozygosity and is more inlnbreeders	Associated with heterozygosity and is morein out breeders	
It is fixable	It is non-fixable	
Selection is very effective as it is fixable	Selection is ineffective as it is non-fixable	
It is the chief cause of transgressive segregation	It is the chief cause of heterosis	

Table 27: Different Types of Gene Actions Involving Two Genes

Component of Genetic Variance	*Specific Symbol*	*Description*	*General Symbol*
Additive	da	The different between AA and aa phenotypicValues	d
	db	The different between BB and bb phenotypic Values	
Dominance	ha	The deviation of Aa phenotype from the average of AA and aa phenotypes	h
	hb	The deviation of Bb phenotype from the average of BB and bb phenotypes	
Epistasis	da x db	Additive x Additive effect due to interaction between AA and BB	i
	da x hb and ha x db	Additive x dominance interaction due to interactions between and AA and Bb and between Aa and BB respectively	j
	ha x hb	Dominance x dominance interaction due to interaction between Aa and Bb	l

Calculation of Variability Parameters using ANOVA Table

The analysis of variance (ANOVA) provides estimates of phenotypic, genotypic and environmental variances, which are used for the estimation of respective coefficient of variation. The phenotypic, genotypic and environmental coefficients of variation are computed as follows (Burton, 1953).

$$PCV = \sqrt{VP}/\overline{X} \times 100, \quad GCV = \sqrt{VG}/\overline{X} \times 100, \quad ECV = \sqrt{VE}/\overline{X} \times 100$$

The phenotype may be described according to a mathematical model to facilitate statistical analysis and interpretation. The phenotypic mean *i.e.* *x* of a given genotype from the trial may be expressed as m,

$$\overline{x} = \mu + g + e + ge$$

where,

$\overline{x}$ = Phenotypic mean

μ = General population mean

g = Effect of genotype

e = Effect of environment

ge = Interaction between genotype and environment

Table 28: ANOVA According to RBD

Source of Variation	*d.f.*	*Expectation of MS*
Replications	r-1	$\sigma_e^2 + g\sigma_r^2$
Genotypes	g-1	$\sigma_e^2 + r\sigma_g^2$
Error	(r-1) (g-1)	σ_e^2
Total	(rg)-1	

Where, g and r are the number of genotypes and replications respectively; and σ_e^2, σ_r^2 and σ_g^2 denote the variances due to error, replications and genotypes respectively.

Phenotypic and genotypic components of variance estimated by applying the formula as suggested by Cochran and Cox (1957).

$$\text{Genotypic variance}\left(\sigma_g^{\ 2}\right) = \frac{\text{MS due to genotypes - MS due to error}}{\text{r}}$$

Phenotypic variance (σ_p^2) = $\sigma_g^2 + \sigma_e^2$ (MS due to error)

where,

r = Number of replications.

Phenotypic (PCV) and Genotypic (GCV) Coefficients of Variations

Phenotypic and genotypic coefficients of variability were computed as per the method suggested by Burton (1953).

$$\text{PCV}(\%) = \frac{\sqrt{\text{Phenotypic variance}}}{\text{General mean}} \times 100$$

or

$$PCV = \frac{\sqrt{\delta^2{}_p}}{\mu} \times 100$$

$$\text{GCV}(\%) = \frac{\sqrt{\text{Genotypic variance}}}{\text{General mean}} \times 100$$

or

$$PCV = \frac{\sqrt{\delta^2{}_g}}{\mu} \times 100 \quad GCV = \frac{\sigma_g}{X} \times 100$$

PCV and GCV were classified as per Robinson *et al.* (1949) *i.e.*0-10 per cent - Low, 10-20 per cent - Moderate and 20 per cent and above – High.

Calculation of Variability Parameters using Segregating Generation Data

Components of variance in F_2 include phenotypic and genotypic variances.

1. Phenotypes variances (σ^2p) = Variance F_2
2. Genotypic variance (σ^2g) = $\sigma^2p - (\sigma^2p_1+\sigma^2p_2+\sigma^2 F_1)/3$

Where, σ^2p_1, σ^2p_2, $\sigma^2 F_1$ are the variance of parents and F_1, respectively.

Heritability

The ratio of genotypic variance to the phenotypic variance or total variance is called as heritability. It is generally expressed in percentage. The estimates of heritability help the plant breeder in selection of elite genotypes from diverse genetic populations. There are two types of heritability, *viz.*, broad sense heritability and narrow sense heritability.

The broad sense heritability (h^2_{bs}) was estimated for all the characters as the ratio of genotypic variance to the total variance as suggested by Hanson *et al.* (1956) as indicated below.

$$h^2{}_{bs} = \frac{\sigma^2{}_g}{\sigma^2{}_p} \times 100$$

Where, σ^2_p is the phenotypic variance and σ^2_g is the genotypic variance of respective characters.

Narrow Sense Heritability

It is the ratio of additive or fixable genetic variance to the total or phenotypic variance. It plays an important role in the selection process in plant breeding.

$h^2_{(ns)} = 1/2D/Vp$ or

$$h^2{}_{ns} = \frac{\sigma^2{}_a}{\sigma^2{}_p} \times 100$$

Where, D or $\sigma^2{}_a$ is the additive genetic variance and Vp or $\sigma^2{}_p$ is the phenotypic variance.

Heritability percentage was categorized as per Robinson *et al.* (1949) as 0-30 per cent - Low, 30-60 per cent - Moderate and 60 per cent and above - High.

Advantages

1. Estimates of heritability are useful in predicting the transmission of character from the parents to their offspring.
2. It helps in the selection of elite types from the mixed parental populations or segregating populations.

3. Narrow sense heritability gives an idea about the additive genetic variance.
4. The broad sense heritability can be estimated from parental as well as hybrid populations.

Disadvantages

1. Estimates of heritability are based on variances and its components and therefore are not statistically very robust and reliable.
2. The variance and its components as compared to means are also inaccurately estimated.
3. Heritability is the property of a specific population in a specific experiment.

Genetic Advance (GA)

Improvement in the mean genotypic value of selected plants over the parental population is known as genetic advance. It is the measure of genetic gain under selection. The grater the genetic variability the higher is the genetic advance and vice- versa.

It was predicted by using the formula provided by Johnson *et al.* (1955).

$$GA = h^2_{(bs)} \times \sigma_p \times k$$

where,

$h^2_{(bs)}$ = Heritability in broad sense

σp = Phenotypic standard deviation of the trait

k = Selection intensity which is 2.06 at 5 per cent selection intensity

Advantages

1. Estimates of genetic advance are based on empirical results and are free from genetical assumption.
2. It is a reliable measure of genetic improvement under selection for polygenic traits.
3. Genetic advance can be estimated both from parental as well as segregating population.
4. Estimates of genetic advance help in understanding the type of gene action involved in the expression of various polygenic characters. High values of genetic advance indicate the presence of additive gene action and low values are indicative of non- additive gene action.

Disadvantages

1. Estimates of genetic advance are based on second order statistics (variance) and therefore are not statistical robust.
2. Estimates of genetic advance are specific to the breeding material; hence it varies according to the breeding material.

Genetic Advance as per cent of Mean (GAM)

It was computed by the formula.

$$\text{GAM (per cent)} = \frac{\text{Genetic advance}}{\text{General mean of the character}} \times 100$$

The genetic advance as per cent mean was categorized as suggested by Johnson *et al.* (1955). 0-10 per cent: Low, 10-20 per cent: Moderate and 20 per cent and above: High.

9

Methods of Breeding of Self-pollinated Crops (Plant Introduction, Mass Selection and Pureline Selection)

The following are the methods of Breeding Autogamous Plants.

1. Introduction
2. Selection
   a) Mass selection
   b) Pure line selection
3. Hybridization and Selection
   i) Inter-Varietal
      a) Pedigree Method
      b) Bulk Method
      c) Single Seed Descent Method
      d) Modified Bulk Method
      e) Mass - Pedigree Method
   ii) Inter-Specific Hybridization
4. Back Cross Method
5. Multiline Varieties

6. Population Approach
7. Hybrids
8. Mutation Breeding
9. Polyploidy Breeding
10. Innovative Techniques

PLANT INTRODUCTION

Plant introduction consists of taking a genotype or a group of genotypes of plants into new environments where they were not being grown before. Introduction may involve new varieties of a crop already grown in the area, wild relatives of the crop species or a totally new crop species. Mostly materials are introduced from other countries or continents. But movement of crop varieties from one environment into another within a country is also introduction. Some examples of within the country introduction are popularization of grape cultivation in Haryana, Introduction of wheat in West Bengal, rice in Punjab *etc.*

Types of Plant Introduction

1. Primary Introduction

When the introduced variety is well suited to the new environment, it is released for commercial cultivation without any alteration in the original genotype, this constitutes primary introduction. Primary introduction is less common, particularly in countries having well organized crop improvement programmes. Introduction of semi dwarf wheat varieties Sonora 64, Lerma Roja and of semi dwarf rice varieties Taichung Native 1 (TN-1), IR- 8 and IR- 36 are some *examples* of primary introductions.

2. Secondary Introduction

The introduced variety may be subjected to selection to isolate a superior variety. Alternatively, it may be hybridized with local varieties to transfer one or few characters from this variety to the local ones these processes are known as secondary introduction. Secondary introduction is much more common than primary introduction. *Examples* of secondary introduction are Kalyan Sona and Sonalika wheat varieties selected from material introduced from CIMMYT, Mexico.

History of Plant Introduction

Crop plants have traveled into many new areas from their centres of origin. This movement of plants occurred with the movement of man. Most of these introductions occurred very early in the history. *For example,* mung mustard, pear, apple and walnut were introduced from the Central Asian Center of Origin into various parts of India. Similarly, sesame, jowar, arhar, Asian cotton and finger millet originated in Africa and traveled to India in the pre-historic period. From this it is clear that plant wealth of various nations is to a large extent the result of plant introductions.

For several centuries A. D., the agencies of plant introduction were invaders, settlers, traders, travelers, explorers and naturalists. The plant introduction was made either knowingly or unknowingly. Muslim invaders introduced in India cherries and grapes from Afghanistan by 1300 A.D. In the 16th century A. D., Portuguese introduced maize, groundnut, chilli, potato, sweet potato, guava pineapple, papaya, cashewnut and tobacco. East India Company brought tea, litchi, and loquat from China. Vegetables like cabbage, cauliflower from the Mediterranean centre. Annatto and mahogany tree species from West Indies in the last quarter of 18th century.

During 19th century, a number of botanical gardens played an important role in plant introduction. The Calcutta botanic garden was established in 1781. The Kew botanic gardens, England arranged introduction of quinine and rubber trees from South America into India. During and after the last part of 19th century various agricultural and horticulture research stations were established in the country. These stations introduced horticulture and agriculture plants independent of each other. There was no co- ordination among these agencies regarding their introduction activities.

Plant Introduction Agencies in India

A centralized plant introduction agency was initiated in 1946 at the Indian Agricultural Research Institute (IARI), New Delhi. The agency began as a *Plant Introduction Scheme* in the Division of Botany and was funded by ICAR. In 1956, during the second five year plan, the scheme was expanded as the *Plant Introduction and Exploration Organisation*. Subsequently in 1961, it was made an independent division in IARI, the *Division of Plant Introduction*. The division was reorganized as *National Bureau of Plant Genetic Resources (NBPGR)* in 1976. The nature of activities and the functions of the bureau have remained the same, but the scope and scale of its activities have increased considerably. The bureau is responsible for the introduction and maintenance of germplasm of agricultural and horticultural plants.

In addition to the National Bureau of Plant Genetic Resources, there are some other agencies concerned with plant introduction. *Forest Research Institute, Dehradun,* has a plant introduction organization which looks after the introduction, maintenance and testing of germplasm of forest trees. *The Botanical Survey of India* was established in 1890; it was responsible for the introduction, testing and maintenance of plant materials of botanical and medicinal interest. But at present, introduction and improvement of medicinal plants is being looked after by NBPGR. The Central Research Institute for various crops, *e.g.*, tea, coffee, sugarcane, potato, tobacco, rice *etc.*, introduce, test and maintain plant materials of their interest. But their activities are coordinated by the NBPGR, which has the ultimate responsibility for introduction activities. Plant material may also be introduced by individual scientists, universities and other research organizations. But all the introductions in India must be routed through the NBPGR, New Delhi.

The National Bureau of Plant Genetic Resources

The bureau has its headquarters at IARI, New Delhi. It has five substations for the testing of introduced plant materials. These substations represent the various climatic zones of India, they are listed below.

1. **Simla:** It is situated in Himachal Pradesh and represents the temperate zone; approximately 2,300 m above sea level.
2. **Jodhpur, Rajasthan:** It represents the arid zone.
3. **Kanya Kumari, Tamil Nadu:** It represents the tropical zone.
4. **Akola, Maharashtra:** It represents the mixed climatic zone. It was recently shifted from Amravati.
5. **Shillong:** For collection of germplasm from North-east India. This part of the country has a large genetic variability for several crop species, *e.g.*, rice, citrus, maize *etc.*

The bureau functions as the central agency for the export and introduction of germplasm of economic importance. The bureau is assisted in its activities by the various Central Research Institutes of ICAR. The activities of the bureau are summarized below:

1. It introduces the required germplasm from its counterparts or other agencies in other countries.
2. It arranges explorations inside and outside the country to collect valuable germplasm.
3. It is responsible for the inspection and quarantine of all the introduced plant materials.
4. Testing, multiplication and maintenance of germplasm obtained through various sources. This may be done by the bureau itself at one of its substations or by one of the concerned Central Institutes of ICAR.
5. Supply, on request, germplasm to various scientists or institutions.
6. Maintenance of records of plant name, variety name, propagating material, special characteristics, source, date and other relevant information about the materials received.
7. Supply germplasm to its counterparts or other agencies in other countries.
8. To publish its exchange and collection lists. An Introduction News Letter containing such lists is being published by the Food and Agriculture Organisation (FAO) since 1957 at irregular intervals. NBPGR has also published some lists, and is in the process of publishing some other catalogues.
9. To set up natural gene sanctuaries of plants where genetic resources are endangered.
10. Improvement of certain plants like medicinal and aromatic plants.

Procedure of Plant Introduction

Introduction consists of the following five steps:

1. Procurement

Any individual or institution can introduce germplasm in India. But all the introductions must be routed through the NBPGR, New Delhi. There are two routes for plant introduction. In first route the individual or the institution makes a direct request to an individual or institution abroad, which has the desired germplasm, to send it through the NBPGR, New Delhi. In second procedure the individual or institute submits his germplasm requirements to the NBPGR with a request for their import.

2. Quarantine

Quarantine means to keep materials in isolation to prevent the spread of diseases *etc.* All the introduced plant propagules are thoroughly inspected for contamination with weeds, diseases and insect pests. Materials that are suspected to be contaminated are fumigated or are given other treatments to get rid of the contamination. If necessary, the materials are grown in isolation for observation of diseases, insect pests and weeds. The entire process is known as quarantine and the rules prescribing them are known as quarantine rules.

3. Cataloguing

When an introduction is received, it is given an entry number. Further, information regarding name of the species, variety, place of origin, adaptation and its various characteristics are recorded. The plant materials are classified into three groups.

1. Exotic collections are given the prefix 'EC'.
2. Indigenous collections are designated as 'IC'.
3. Indigenous wild collections are marked as 'IW'.

4. Evaluation

To assess the potential of new introductions, their performance is evaluated at different substations of the Bureau. In case of those crops for which Central Research Institutes are functioning (*e.g.*, rice, sugarcane, potato, tobacco *etc.*), the introduced materials are evaluated and maintained by these institutes. The resistance to diseases and pests is evaluated under environments favouring heavy attacks by them.

5. Multiplication and Distribution

Promising introductions or selections from the introductions may be increased and released as varieties after the necessary trials. Most of the introductions, however, are characterized for desirable traits and are maintained for future use. Such materials are used in crossing programmes and are readily supplied by the bureau on request.

Acclimatization

Generally, the introduced varieties perform poorly because they are often not adapted to the new environment. Sometimes, the performance of a variety in the new environment improves with the number of generations grown there. The process that leads to the adaptation of a variety to a new environment is known as acclimatization. Acclimatization is brought about by a faster multiplication of those genotypes (present in the original population) that are better adapted to the new environment. Thus acclimatization is essentially natural selection. Variability must be present in the original population for acclimatization to occur. Therefore, land varieties are likely to get acclimatized, while pure lines are not likely to do so.

The extent of acclimatization is determined by (1) the mode of pollination, (2) the range of genetic variability present in the original population, (3) the duration of life- cycle of the crop and (4) the intensity of stress in the new area. Cross-pollination leads to far greater gene recombinations than self-pollination. As a result cross-pollination is much more helpful in acclimatization than self-pollination.

Purpose of Plant Introduction

The main purpose of plant introduction is to improve the plant wealth of the country. The chief objectives of plant introduction may be grouped as follows:

- ☆ **To Obtain an Entirely New Crop Plant:** Plant introductions may provide an entirely new crop species. Many of our important crops, *e.g.*, maize, potato, tomato, tobacco, *etc.*, are introductions. Some recently introduced crops are soybean, gobhi sarson, oil palm *etc.*
- ☆ **To Serve as New Varieties:** Sometimes introductions are directly released as superior commercial varieties. The Maxican semidwarf wheat varieties Sonora 64 and Lerma Rojo, semi dwarf rice varieties TN 1, IR- 8 and IR -36 are more recent *examples* of this type.
- ☆ **To be used in Crop Improvement:** Often the introduced material is used for hybridization with local varieties to develop improved varieties. Pusa Ruby tomato was derived from a cross between Meeruty and Sioux, an introduction from U.S.A.
- ☆ **To Save the Crop from Diseases and Pests:** Sometimes a crop is introduced into a new area to protect it from diseases and pests. Coffee was introduced in South America from Africa to prevent losses from leaf rust. Hevea rubber, on the other hand, was brought to Malaya from South America to protect it from a leaf disease.
- ☆ **For Scientific Studies:** Collections of plants have been used for studies on biosystematics, evolution and origin of plant species. N.I. Vavilov developed the concept of centres of origin and that of homologous series in variation from the study of a vast collection of plant types.
- ☆ **For Aesthetic Value:** Ornamentals, shrubs and lawn grasses are introduced to satisfy the finer sensibilities of man. These plants are used for decoration and are of great value in social life.

- ✰ **Varieties selected from Introductions:** Many varieties have been developed through selection from introductions. Two varieties of wheat, Kalyan Sona and Sonalika, were selected from introductions from CIMMYT, Mexico.
- ✰ **Varieties developed through Hybridization:** Introductions have contributed immensely to the development of crop varieties through hybridization. All the semi dwarf wheat varieties are derived from crosses with Mexican semi-dwarf wheat. All but few semi dwarf rice varieties possess the dwarfing gene from *dee- geo-woo -gen* through either TN1 or IR 8. Thus almost all these semi- dwarf wheat and rice varieties have been developed from crosses involving introductions. All the sugarcane varieties have been derived from the introduced noble canes.

Other examples of varieties developed through hybridization with introductions are Pusa Ruby tomato obtained from a cross between Meeruti and Sioux; Pusa Early Dwarf Tomato derived from the cross Meeruti x Red Cloud; Pusa Kesar carrot, Pusa Kanchan turnip *etc.*

Merits of Plant Introduction

1. It provides entirely new crop plants.
2. It provides superior varieties either directly or after selection and hybridization.
3. Introduction and exploration are the only feasible means of collecting germplasm and to protect variability from genetic erosion.
4. It is very quick and economical method of crop improvement, particularly when the introductions are released as varieties either directly or after a simple selection.
5. Plants may be introduced in new disease free areas to protect them from damage, *e.g.*, coffee and rubber.

Demerits of Plant Introduction

Introduction also responsible for introduction of below mentioned problematic weeds, pathogen, pests, *etc.*

1. Weeds: *Argemone maxicana, Eichornia crassipes, Phylaris minor etc.*
2. Diseases:
   - ✰ Late blight of potato from France (in 1883).
   - ✰ Flag smut of wheat- Australia.
   - ✰ Coffee rust- Ceylon (1886).
   - ✰ Bunchy top of banana- Ceylon (1940), *etc.*
3. Insect Pests:
   - ✰ Potato tuber moth- Italy (1900).
   - ✰ Woolly aphis of apple.

- Fluted scale of citrus, *etc.*

4. Ornamental turn Weeds
   - Water hyacinth
   - Lantana camera, *etc.*
5. Threat to Ecological Balance
   - Eucalyptus spp. – Australia- cause Rapid depletion of sub-soil water reserves.

SELECTION

- **'Differential reproduction rate of different genotypes'**
- Selection is basic to any crop improvement. Isolation of desirable plant types from the population is known as selection. It is one of the two fundamental steps of any breeding programme *viz.*, creation of variation and selection.
- There are two agencies involved in carrying out selection *viz.*, Natural Selection and Artificial Selection. Though both may complement each other in some cases, they are mostly opposite in direction, since their aims are different under the two conditions (nature and domestication).
- The effectiveness of selection primarily depends upon the degree to which phenotype reflects the genotype.
- Before domestication, crop species were subjected to natural selection. The basic for natural selection was adaptation to the prevailing environment. After domestication man has knowingly or unknowingly practiced some selection. Thus crop species under domestication were exposed to both natural and artificial selection. For a long period, natural selection played an important role than selection by man. But in modern plant breeding methods natural selection is of little importance and artificial selection plays an important role.

History of Selection

- Selection was practiced by farmers from ancient times.
- During 16th century Van Mons in Belgium, Andrew knight in England and Cooper in USA practiced selection in crop plants and released many varieties.
- Le coutier, a farmer of island of New Jersey published his results on selection in wheat in the year 1843. He concluded that progenies from single plants were more uniform.
- During the same period Patrick shireff, a Scotsman practiced selection in wheat and oats and developed some valuable varieties.
- During 1857 Hallet in England practiced single plant selection in wheat, oats and barley and developed several commercial varieties.

- ☆ About this time **Vilmorin** proposed individual plant selection based on progeny testing. This method successfully improved the sugar content in sugar beet. His method was called as *Vilmorins Isolation Principle*. He emphasised that the real value of a plant can be known only by studying the progeny produced by it. This method was successful in sugar beet but not in wheat. This shows the in-effectiveness of selection in cross-pollinated crops. Today progeny test is the basic step in every breeding method.

Basic Principles of Selection

Notwithstanding the highly complex genetic situation imposed by linkage and epistasis, there are just three basic principles of selection (Walker,1969).

1. **Selection operates on existing variability**: The main function of the selection exercise is to discriminate between individuals. This is possible only when sufficient variation is present in the material subjected to selection pressure. Thus, selection acts on the existing variation it cannot create new variation.
2. **Selection acts only through heritable differences**: only the selected individuals are permitted to contribute to the next generation/progenies. Therefore, should there be greater influence of non- heritable agencies on the individuals selected; the parent-progeny correlation will be greatly vitiated. Hence the variation among individuals to be selected must be genetic in nature, since it is the genetic variation that tends to close the gap between phenotype and genotype. Environmental variability cannot be of any use under selection.
3. **Selection works because some individuals are favoured in reproduction at the expense of others**: As a consequence of its past evolutionary history and breeding structure, a population or a crop consists of highly genetically variable individuals with regards to such diverse phenomena as differential viability, differential maturity, differences in mating tendencies, fecundity, and duration of reproductive capacity. Hence some individuals tend to become superior to others for some or other traits desirable under domestication. These superior individuals are retained for reproduction while others discarded under selection.

Selection has two basic characteristics *viz.*

1. Selection is effective for heritable differences only.
2. Selection does not create any new variation. It only utilizes the variation already present in a population.

Selection Intensity

Percentage of plants selected, to be advanced to next generation, from a population.

Selection Intensity (I): Is the amount of selection applied expressed as the proportion of the population favoured (selected). The selection intensity is inversely proportional to the percentage proportion selected (PS), as reflected in Table 29.

(i) = S/ δP

Table 29: Relationship between Selection Intensity and Proportion of Population Selected

Selection intensity (I or K) (i)	2.64	2.42	2.06	1.76	1.40	1.16	0.97	0.80	0.34	0
Per cent selected (PS) (q/n * 100)	1%	2%	5%	10%	20%	30%	40%	50%	80%	100%

Thus, larger the size of I, more stringent is the selection pressure (hence low fraction is selected) and vice versa. Then, no selection means all the members of a population are allowed to reproduce (I=0, PS=100 per cent), and zero selection means the whole population is rejected (PS=0). However, in real selection experiments, as the desired alleles become preponderant after each cycle of selection, I, is also changed.

The I is of the greatest consequence is bringing about changes in the gene frequency under selection. However, since the latter does not mean undue loss of desirable alleles, or undue load of population size, the choice of an arbitrary value of I may be hazardous in a plant breeding programme. The small size of I (*i.e.* low selection pressure) may cause a large population size to be handled in the next generation, which will unnecessarily be taxing on time and resources. On the other hand, a large size of I (high selection pressure) might cause allelic erosion due to genetic drift (*i.e.* changes in gene frequencies due to sampling error, or small sample size under selection in a finite population not due to genetic causes). Therefore, an appropriate level of I should be chosen based upon the range of variability present in the population subjected to selection.

The I=2.06 to 1.76 (*i.e.* around 5-10 per cent of individuals selected) has generally been found appropriate in plant populations. However, the limit of selection intensity is set by two factors: (i) population size, and (ii) inbreeding. Under natural selection, selection intensity is expressed as the relative number of offsprings produced by different genotypes, and is termed as *selection coefficient*.

Selection Differential

Difference between the mean of the population and mean of the selected individuals. Expressed in terms of standard deviation and is designated as 'S'. S is the average superiority of selected individuals over the mean of population of their origin. It is considered in the same parental generation before selection is made. An arbitrary culling level, k (*i.e.* I) is fixed for a trait and individuals beyond that level are selected. The average of all such selected individuals can be designated by X. Then the mean of selected individuals (μ) exceeds the parental population mean by the measure of S.

That is S=X-μ. Therefore, wider the phenotypic variability (*i.e.* phenotypic variance, and phenotypic standard deviation, that measures variability), greater is the possibility of S being large.

Heritability

In crop improvement, only the genetic component of variation is important since only this component is transmitted to the next generation. The ratio of genetic variance to the total variance *i.e.* phenotypic variance is known as heritability.

$H_{(bs)} = Vg/Vp$

$Vp = Vg+Ve$

where,

Vp = Phenotypic variance

Vg = Genotypic variance

Ve = Error variance or environmental variance

Heritability estimated from the above formula is known as the broad sense heritability. This is valid when homozygous lines are studied. But when segregating generations are studied genotypic variance consists of (a) additive variance (b) dominance variance (c) and variance due to epistasis.

Dominance variance is important when we are dealing with hybrids *i.e.* F_1 generations. In self pollinated crops we release varieties only after making them homozygous lines. Hence additive variance is more important in such cases. The proportion of additive genetic variance to the total variance is known as narrow sense heritability.

$H_{(ns)} = VA/VP$

If heritability is very high for any character it can be improved. Improvement of characters with low heritability is very difficult.

Genetic Advance under Selection

Normally selection is practiced based on the phenotype of the individual plant. The phenotype in turn is the result of joint action of genotype and environment *i.e.*,

$VP = VG + VE$

where,

P = Phenotype

G = Genotype

E = Environment

It is the improvement in the mean genotypic value of the selected families over the base population is known as genetic advance under selection.

Genetic Advance under Selection Depends upon

a) Genetic variability among different plants or families in the base population.

b) The heritability of the character under selection.

c) The intensity of selection *i.e.*, the proportion of plants or families selected.

Genetic advance under selection may calculate as follows:

GS = (K) (σP) (H)

where,

GS = Genetic advance under selection

K = Selection intensity

σP = Phenotypic standard deviation of the base population

H = Heritability of the character under selection.

MASS SELECTION

A large number of plants of similar phenotype are selected and their seeds are mixed to constitute the new variety.

Mass selection is an example of selection from a biologically variable population in which differences are genetic in origin. The Danish biologist, **W. Johansen**, is credited with developing the basis for mass selection in 1903. Mass selection is often described as the oldest method of breeding self-pollinated plant species. However, this by no means makes the procedure outdated. As an ancient art, farmers saved seed from desirable plants for planting the next season's crop, a practice that is still common in the agriculture of many developing countries. This method of selection is applicable to both self- and cross-pollinatedspecies.

Applications

1. It may be used to maintain the purity of an existing cultivar that has become contaminated, or is segregating. The off-types are simply rogued out of the population, and the rest of the material bulked. Existing cultivars become contaminated over the years by natural crosses (*e.g.*, out crossing, mutation) or by human error (*e.g.*, inadvertent seed mixture during harvesting or processing stages of crop production).
2. It can also be used to develop a cultivar from a base population created by hybridization, using the procedure described next.
3. It may be used to preserve the identity of an established cultivar or soon-to-be-released new cultivar. The breeder selects several hundreds (200–300) of plants (or heads) and plants them in individual rows for comparison. Rows showing significant phenotypic differences from the other rows are discarded, while the remainder is bulked as breeder seed. Prior to bulking,

sample plants or heads are taken from each row and kept for future use in reproducing the original cultivar.

4. When a new crop is introduced into a new production region, the breeder may adapt it to the new region by selecting for key factors needed for successful production (*e.g.*, maturity). This, hence, becomes a way of improving the new cultivar for the new production region.

Procedure for Evolving Variety by Mass Selection (Without Progeny Test)

First Year

Large number of phenotypically similar plants having desirable characters are selected. The number may vary from few hundred to few thousand. The seeds from the selected plants are composited to raise the next generation.

Second Year

Composited seed planted in a preliminary field trial along with standard checks.

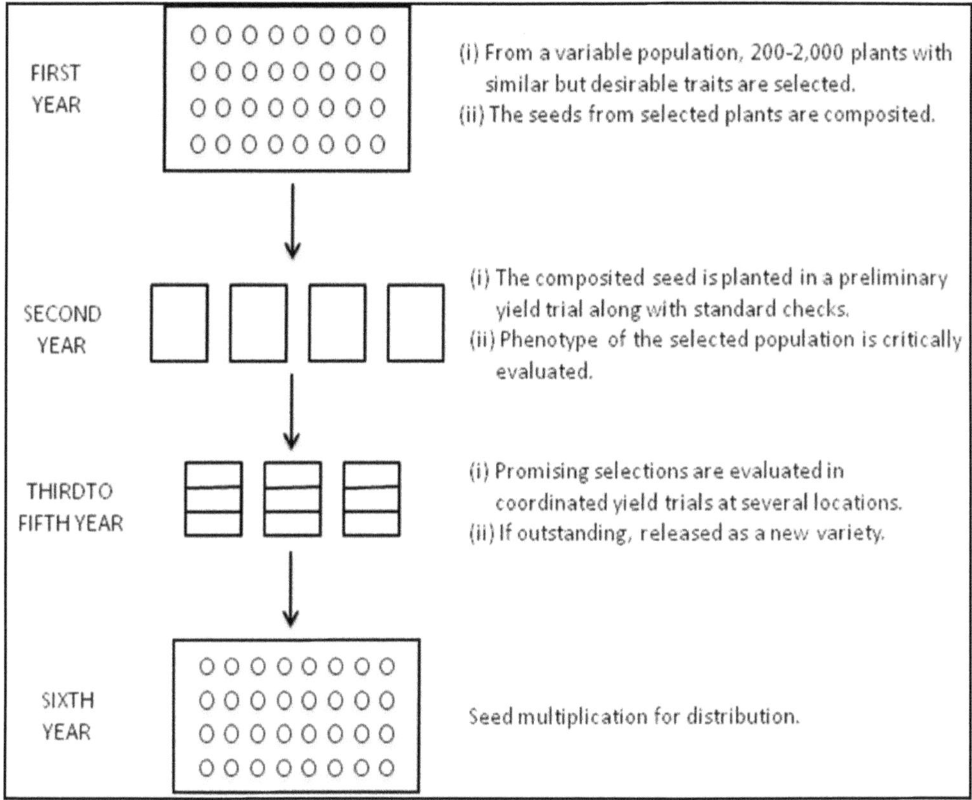

Figure 15: Mass Selection in Self-pollinated Crops as Used to Develop New Varities. For maintaining the purity of pureline varieties, operations of the first year may be repeated every year or after every few years.

The variety from which the selection was made should also be included as check. Phenotypic characteristics of the variety are critically examined and evaluated.

Third to Sixth Year

The variety is evaluated in coordinated yield trials at several locations. It is evaluated in an initial evaluation trial (IET) for one year. If found superior it is promoted to main yield trials for 2 or 3 years.

Seventh Year

If the variety is proved superior in main yield trials it is multiplied and released after giving a suitable name.

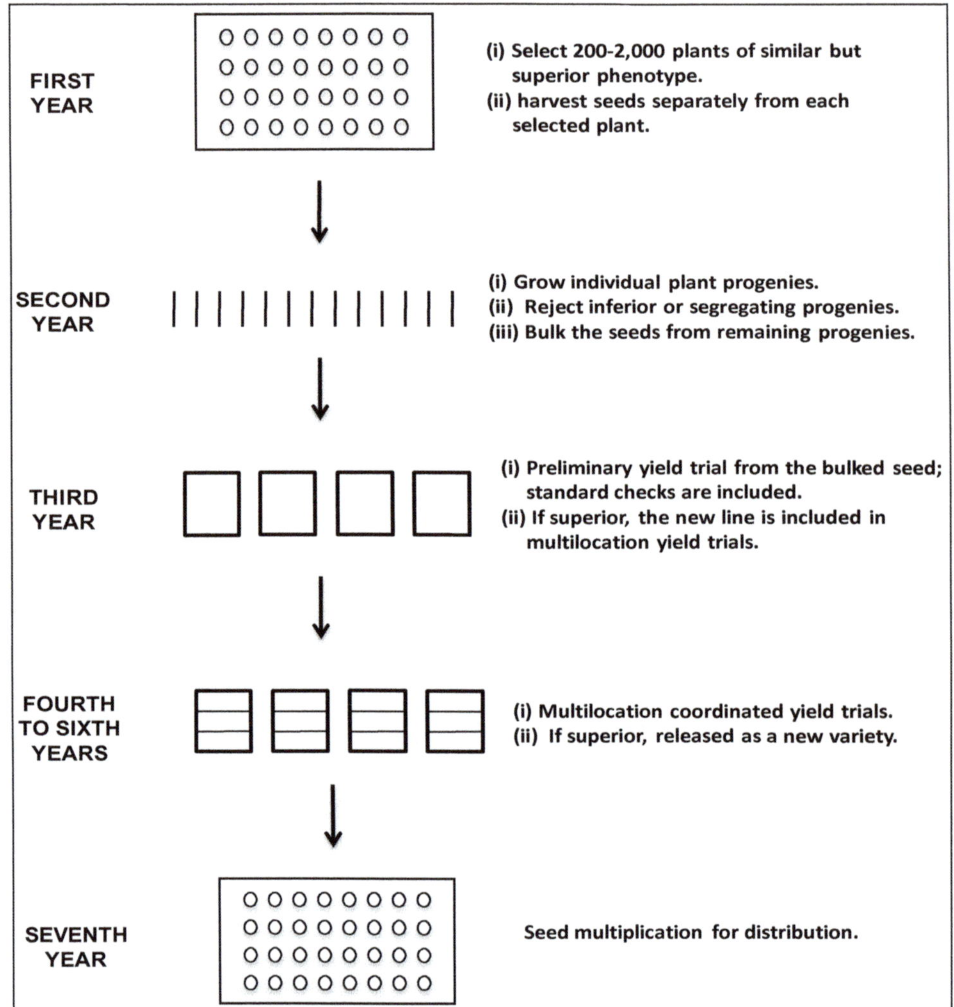

Figure 16: Mass Selection in Self-pollinated Crops Coupled with Progeny Test. Thie method is more useful that the one outlined before. It is commonly used for maintaining the purity of pureline varieties (steps of year 1 and 2 only).

Modification of Mass Selection

Mass selection is used for improving a local variety. Large number of plants are selected (I year) and individual plant progenies are raised (II year). Inferior, segregating progenies are rejected. Uniform, superior rows are selected and the seed is bulked. Preliminary yield trials are conducted in third year. Fourth to seventh year multilocation tests are conducted and seed is multiplied in eight year and distributed in ninth year. Many other modifications also are followed depending on the availability of time and purpose for which it is used.

Merits of Mass Selection

1. Can be practiced both in self and cross-pollinated crops.
2. The varieties developed through mass selection are more widely adapted than pure lines.
3. It retains considerable variability and hence further improvement is possible in future by selection.
4. Helps in preservation of land races.
5. Useful for purification of pureline varieties.
6. Improvement of characters governed by few genes with high heritability is possible.
7. Less time consuming and less expensive.

Demerits of Mass Selection

1. Varieties are not uniform.
2. Since no progeny test is done, the genotype of the selected plant is not known.
3. Since selection is based on phenotype and no control over pollination the improvement brought about is not permanent. Hence, the process of mass selection has to be repeated now and then.
4. Characters which are governed by large number of genes with low heritability cannot be improved.
5. It cannot create any new genotype but utilizes existing genetic variability.

Achievements

Mass selection must have been used by pre historic man to develop present day cultivated cross from their wild parents. It was also used extensively before pureline selection came into existence.

Cotton: American Cotton

Groundnut: TMV- 1, TMV-2 *etc.*

Bajra: Pusa moti, Bajapuri, Jamnagar Giant, AF3, *etc.*

Sorghum: R.S. 1

Rice: SLO 13, MTU- 15, *etc.*

Potato: K122, *etc.*

THE PURE LINE THEORY

A pureline is a progeny of a single homozygous plant of a self-pollinated species. All the plants of a pureline have the same genotype. The phenotypic differences within apureline is due to environment. Therefore variation within a pureline is not heritable. Hence selection in a pureline is not effective.

The concept of pureline was proposed by ***Johannsen*** in **1903** on the basis of his studies with *Princess* variety of common beans (*Phaseolus vulgaris*). From a commercial seed lot he selected seeds of different sizes and grew them separately. The progenies differed in seed size. Progenies from larger seeds produced larger seeds than those obtained from smaller seeds. This clearly showed that the variation in seed size in the commercial seed lot of princess variety had a genetic base. As a result selection for seed size was effective.

Johannsen, further studied 19 lines, each line was a progeny of a single seed from the original lot. He discovered that each line showed a characterize mean seed weight, ranging from 640mg in Line No 1 to 350 mg in line No 19. The seed

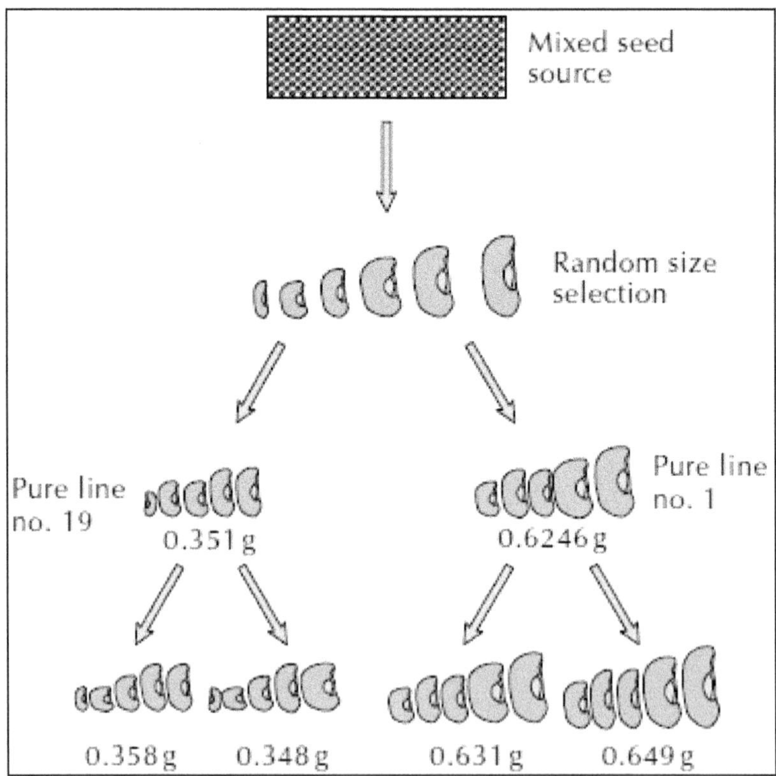

Figure 17: Development of Pure Line.

size within a line showed some variation, which was much smaller than that in the original commercial seed lot. Johannsen postulated that the original seed lot was a mixture of purelines. Thus each of the 19 lines represented a pureline, and the variation in seed size within each of the pure lines had no genetic basis and was entirely due to environment.

The two main conclusions from the Johannsens' experiment are:

1. A self-fertilized population consists of a mixture of several homozygous genotypes. Variation in such a population has a genetic component, and therefore selection is effective.
2. Each individual plant progeny selected from a self -fertilized population consists of homozygous plants of identical genotype. Such a progeny is known as pureline. The variation within a pureline is purely environmental and, as a result, selection within a pureline is ineffective.

Characters of Purelines

1. All the plants within a pureline have the same genotype
2. The variation within a pureline is environmental and non-heritable
3. Purelines are stable

Genetic Basis of Pure Line

Self pollination increases homozygosity with a corresponding decrease in heterozygosity. The effect of self pollination on homogygosity and heterozygosity is illustrated by taking an individual heterozygous for (*Aa*) a single gene (Table 30).

Table 30: The Effect of Selfing on Homozygosity and Heterozygosity

No of Generations of Selfing	*Frequency Per cent*			*Frequency Per cent*	
	AA	*Aa*	*aa*	*Homozygosity*	*Heterozygosity*
0	0	100	0	0	100
1	25	50	25	50	50
2	37.5	25	37.5	75	25
3	43.75	12.5	43.75	87.5	12.5
4	46.875	6.25	46.875	93.73	6.25
5	48.437	3.125	48.437	96.874	3.125
6	49.218	1.562	49.218	98.436	1.562
7	49.608	0.781	49.608	99.216	0.781
8	49.803	0.39	49.803	99.606	0.39

Proportion of completely homozygous plants in the population

$= [(2^m - 1)/2^m]^n$

where,

m = No. of generations of self pollination

n = No. of genes segregating.

Suppose an individual heterozygous for a single gene (*Aa*) and the successive generations derived from it are subjected to self- pollination. Every generation of self -pollination will reduce the frequency of heterozygote *Aa* to 50 per cent of that in the previous generation.

There is a corresponding increase in the frequency of the two homozygotes *AA* and *aa*. As a result, after 10 generations of selfing, virtually all the plants in the population would be homozygous, *i.e.*, *AA* and *aa*. On the other hand, the frequency of heterozygote *Aa* would be only 0.097 per cent, which is negligible. It is assumed here that the three genotypes *AA, Aa* and *aa* have equal survival. If there is unequal survival, it may increase or decrease the rate at which homozygosity is achieved. If *Aa* is favoured, the rate of increase in homozygosity would be lower than expected. But if *Aa* is selected against, hom ozygosity would increase at a faster rate than expected.

Applications

1. Cultivars for mechanized production that must meet a certain specification for uniform operation by farm machines (*e.g.*, uniform maturity, uniform height for location of economic part).
2. Cultivars developed for a discriminating market that puts a premium on visual appeal (*e.g.*, uniform shape, size).
3. Cultivars for the processing market (*e.g.*, demand for certain canning qualities, texture).
4. Advancing "sports" that appear in a population (*e.g.*, a mutant flower for ornamental use).
5. Improving newly domesticated crops that have some variability (*e.g.*, NP 4, NP 52 in Wheat)
6. The pure-line selection method is also an integral part of other breeding methods such as pedigree selection and bulk population selection.

General Procedure for Evolving a Variety by Pure Line Selection

First Year

A large number of plants (200- 3000) which are superior to the rest are selected from a local variety or mixed population and harvested separately (in some cases individual heads or stems may be selected). The number of plants to be selected depends upon the breeder's discretion but should be as large as possible in view of the available time, land, funds, labour *etc.* It is advisable to select for easily observable characters such as flowering, maturity, disease resistance, plant height *etc.*

Second Year

Progenies of individual plants selected in 1st year are grown separately with proper spacing (plant to row or head to row). The progenies are evaluated by taking elaborate data on visual characters such as plant height, duration, grain type, ear characters besides yield. The number of progenies should be reduced as much as possible. Disease epiphytotics may be created to test the progenies for disease resistance, poor, weak, diseased, insect attacked and segregating progenies are rejected. The superior progenies are harvested separately. If necessary the process may be repeated for one or more years.

Third Year

The selected progenies, now called as cultures are grown in replicated trial for critical evaluation of yield *etc.* The best local variety is used as a check and should be grown at regular intervals, after every 15 or 20 cultures for comparison. This is known as preliminary yield trial. Superior cultures based on observable characters and yields are selected. The number is drastically reduced.

Fourth and Sixth Years

The superior cultures are tested against the local checks in yield trials. Observations are recorded on many characters like diseases resistance, days to flower, days to maturity, height of the plant ear characters, test weight and yield. The data is subjected to statistical analysis to identify really superior cultures. If necessary the trials may be extended for one more year or season. Inferior culture are rejected and a few (4-5) promising cultures are selected.

Seventh Year

The best progeny identified earlier is multiplied, named and released as a variety for official release of any variety (approval from the variety releasing committee of the state or central is necessary).

Advantage of Pureline Selection

1. The purelines are extremely uniform since all the plants in the variety will have the same genotype.
2. Attractive and liked by the farmers and consumers.
3. Purelines are stable and long test for many years.
4. Due to its extreme uniformity the variety can be easily identified in seed certification programmes.

Limitations or Disadvantages of Pureline Selection

1. New genotypes are not created by pureline selection.
2. Improvement is limited to the isolation of the best genotype present in population. No more improvement is possible after isolation of the best available genotype in the population.
3. Selection of purelines require great skill and familiarity with the crop.
4. Difficult to detect small differences that exist between cultures.
5. The breeder has to devote more time.

6. Pure lines have limited adaptability hence can be recommended for cultivation in limited area only.

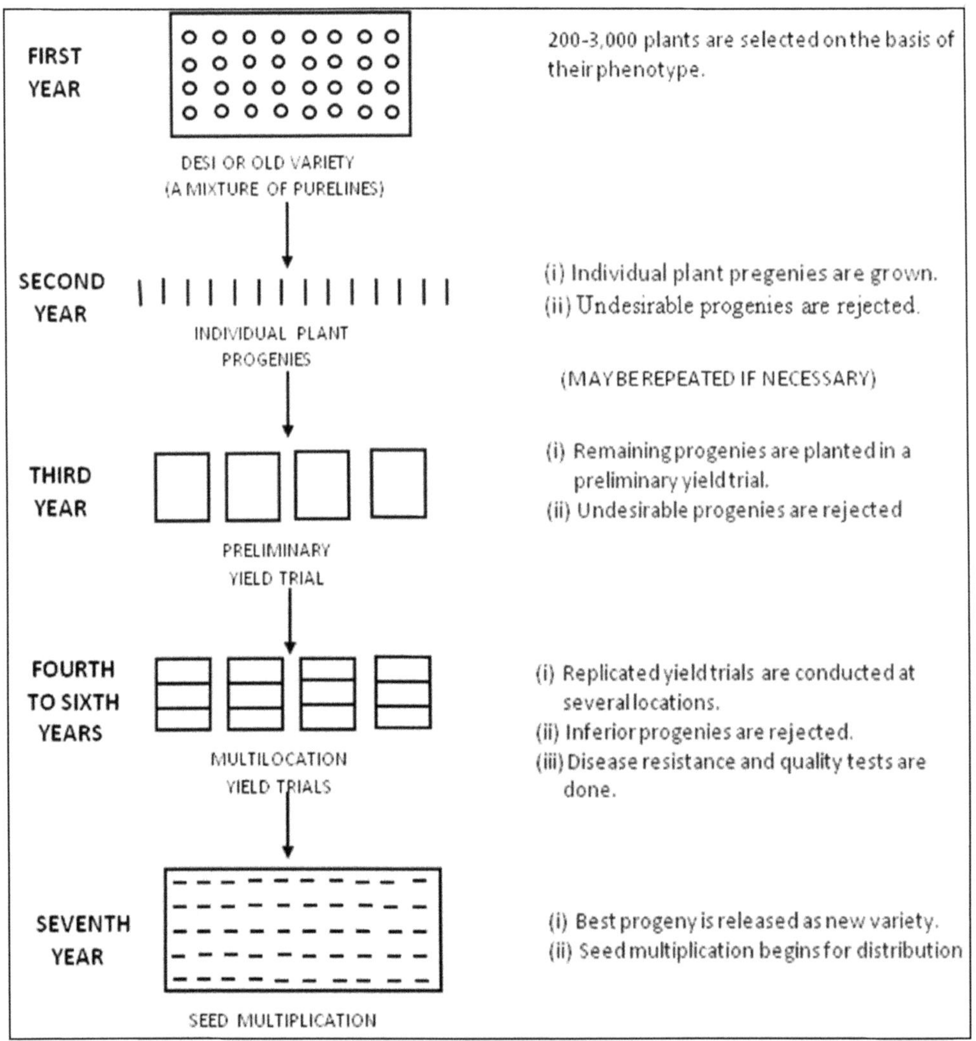

Figure 18: Schematic Representation of Pureline Selection in Self-pollinated Crops.

Achievements

Several varieties developed by pureline selection were released in many crops. Some examples are given below:

Rice: MTU- 1, MTU- 3, MTU- 7, BCP- 1, ADT-1, 3, 5, and 10, *etc.*

Sorghum: G 1 and 2, M 1 and 2, OO 1, 4 and 5, *etc.*

Groundnut: TMV 3, 4, 7, 8 and Kadiri 71-1, *etc.*

Red gram: TM-1, ST- 1, *etc.*

Chillies : G1 and G2, *etc.*

Ragi : AKP 1 to 7, *etc.*

Table 31: Difference between Pure Line and Mass Selection

Sl.No.	*Pure Line Selection*	*Mass Selection*
1.	The new variety is a pure line	The new variety is a mixture of pure lines.
2.	The new variety is highly uniform. Infact, the variation within a pure linevariety is purely environmental.	The variety has genetic variation of quantitative characters, although it is relatively uniform in general appearance.
3.	The selected plants are subjected toprogeny test.	Progeny test is generally not carried out.
4.	The variety is generally the best pure line present in the original population. The pure line selection brings about the greatest improvement over the original variety.	The variety is inferior to the best pureline because most of the pure lines included in it will be inferior to the best pure line.
5.	Generally, a pure line variety is expected to have narrower adaptation and lower stability in performance than a mixture of pure lines.	Usually the variety has a wider adaptation and greater stability than a pure line variety.
6.	The plants are selected for the desirability. It is not necessary they should have a similar phenotype.	The selected plants have to be similar in phenotype since their seeds are mixed to make up the new variety.
7.	It is more demanding because careful progeny tests and yield trials have to be conducted.	If a large number of plants are selected, expensive yield trials are not necessary. Thus it is less demanding on the breeder.

10

Hybridization Methods

The mating or crossing of two plants or lines of dissimilar genotype is known as hybridization. In plants, crossing is done by placing pollen grains from one genotype, the male parent, on to the stigma of flowers of the other genotype, the female parent. It is essential to prevent self- pollination as well as chance cross- pollination in the flowers of the female parent. At the same time, it must be ensured that the pollen from desired male parent reaches the stigma of female flowers for successful fertilization. The seeds as well as the progeny resulting from the hybridization are known as hybrid or F_1. The progeny of F_1, obtained by selfing or intermating of F_1 plants, and the subsequent generations are termed as segregating generations. The term cross is often used to denote the products of hybridization, *i.e.* the F_1 as well as the segregating generations.

History

- 700 BC- Babylonians and Assyrians- hand pollinated date palm for metaxenic effects of pollen (the effect of pollen on the maternal tissue of the fruit).
- 1694- Camararious- identified sex in plants.
- 1717- Fairchild- produced first inter-specific hybrid
- 1759-1835- Joseph Koelreuter- attempted many crosses in tobacco.
- Goss, Sargaret, Gaertnor, and Naudin observed uniformity in F_1, dominance in F_1 and segregation in F_2 generations.
- 1865- Mendel- laws of inheritance of qualitative traits.

Today hybridization is most common method of crop improvement, and the vast majority of crop varieties have resulted from hybridization.

Objectives of Hybridization

The chief objective of hybridization is to create genetic variation. When two genotypically different plants are crossed, the genes from both the parents are brought together in F_1. segregation and recombination produce many new gene combinations in F_2 and the later generations, *i.e.* the segregating generations. The degree of variation produced in the segregating generations would, therefore, depend on the number of heterozygous genes in the F_1. This still, in turn, depend upon the number of the genes for which the two parents differ. If the two parents are closely related, they are likely to differ for a few genes only. But if they are not related, or are distantly related, they may differ for several, even a few hundred, genes. However, it is not likely that the two parents will ever differ for all their genes. Therefore, when it is said that the F_1 is 100 per cent heterozygous, it has reference only to those genes for which the two parents differ.

The aim of hybridization may be the transfer of one or few qualitative characters, the improvement in one or more quantitative characters, or use of F_1 as a hybrid variety. These objectives are briefly discussed below.

Combination Breeding

The main aim of combination breeding is the transfer of one or more characters into a single variety from other varieties. These characters may be governed by oligogenes or polygenes. The intensity of the character in the new variety is either comparable to or, more generally, lower that in the parent variety from which it was transferred. In this approach, increase in the yield of a variety is obtained by correcting the weaknesses in the yield contributing traits, *e.g.*, tiller number, grains per spike; test weight is that for disease resistance. The backcross method of breeding was designed for combination breeding, and often pedigree method also fulfils the same purpose. In combination breeding, the genetic divergence between parents is not the major consideration. What is important is that one of the parents must have in a sufficient intensity the character(s) under transfer, while the other parent is generally a popular variety.

Transgressive Breeding

Transgressive breeding aims at improving yield or its contributing characters through transgressive segregation. Transgressive segregation is the production of plants in an F_2 generation that are superior to both the parents for one or more characters. Such plants are produced by an accumulation of plus or favourable genes from both the parents as a must combine well with each other, and should preferably be genetically diverse, *i.e.*, quite different. This way, each parent is expected to contribute different plus genes which when brought together by recombination give rise transgressive segregant. As a result, the intensity of character in the transgressive segregant, *i.e.*, the new variety, is greater than that in either of the parents. The pedigree method of breeding and its modifications, particularly the population approach, are designed for the production of transgressive segregants.

Hybrid Varieties

In most self- pollinated crops, F_1 is more vigorous and higher yielding than the parents. Wherever it is commercially feasible, F_1 may be used directly as a variety. In such cases, it is important that the two parents should produce an outstanding F_1.

Types of Hybridization

The plants or lines involved in hybridization may belong to the same variety, different varieties of the same species, different species of the same genus or species from different genera. Based on the taxonomic relationship of the two parents, hybridization may be classified into two broad groups:

1. Inter-varietal
2. Distant hybridization

Inter-varietal Hybridization

The parents involved in hybridization belong to the same species; they may be two strains, varieties or races of the same spicies. It is also known as intraspecific hyrbidization. In crop improvement programmes, inter-varietal hybridization is the most commonly used. In fact, it is so common that it may often appear to be the only form of hybridization used in crop improvement. An example would be crossing of two varieties of wheat, rice or some other crop. The inter-varietal crosses may be simple or complex depending upon the number of parents involved.

Simple Cross

In a simple cross, two parents are crossed to produce the F_1. The F_1 is selfed to produce F_2 or is used in a backcross programme, *e.g.*, A X B $\rightarrow$ F_1 (A X B)

Complex Cross

More than two parents are crossed to produce the hybrid, which is then used to produce F_2 or is used in a backcross. Such a cross is also known as convergent cross because this crossing programme aims at converging, *i.e.*, bringing together, genes from several parents into a single hybrid (Figure 19).

Crop improvement progresses, the crop varieties would accumulate more and more favourable genes. This would lead to greater similarities between even unrelated varieties. In view of this, it may be expected that in future complex crosses would become more and more important. In breeding of highly improved self-pollinated crops like wheat and rice, complex crosses are a common practice today. Complex crosses would become routine in near future in the improvement of other self-pollinated crops with the progress in the level of their improvement.

Types of Hybrids

Depending on the parents, there are a number of different types of hybrids

- ☆ **Single cross hybrids**: Result from the cross between two true breeding organisms and produces an F_1 generation called an F_1 hybrid. The cross between two different homozygous lines produces an F_1 hybrid that is

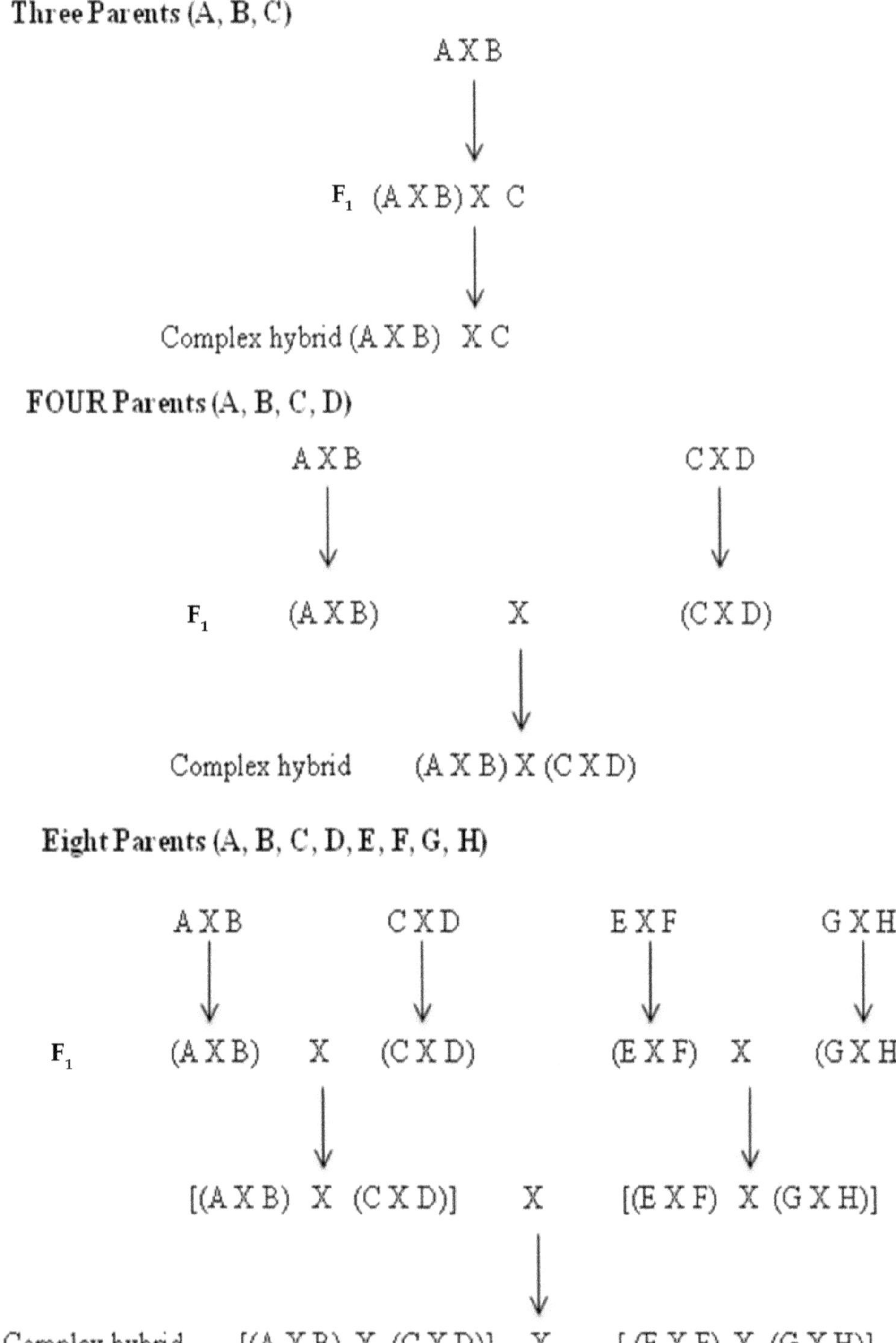

Figure 19: Types of Complex Crosses Based on Number of Parents Involved.

heterozygous; having two alleles, one contributed by each parent and typically one is dominant and the other recessive. The F_1 generation is also phenotypically homogeneous, producing offspring that are all similar to each other.

- ✰ ***Three-way hybrids***: Three parents are involved in a three-way cross. The female of a three-way hybrid is a single-cross hybrid, while the male is an inbred line.
- ✰ **Double-cross hybrids**: In this cross, both parents are single-cross hybrids.
- ✰ ***Top-cross***: In this case, one of the parents is an open-pollinated variety and the other is a single-cross hybrid or an inbred line.
- ✰ **Triple cross hybrids**: Result from the crossing of two different three-way cross hybrids.
- ✰ **Population hybrids**: Result from the crossing of plants in a population with another population. These include crosses between organisms such as interspecific hybrids or crosses between different races.
- ✰ **Poly Cross Hybrids:** Open pollination (assumed to be random mating) in isolation among a number of selected genotypes; also progeny from such a cross.

Distant/Wide Hybridization

- ✰ When crosses are made between two different species (inter-specific/ intra-generic) or between two different genera (inter-generic) of the same family. An efficient method of transferring desirable genes into cultivated plants from their related species.

History

- ✰ Thomas Fairchild 1717 was the first man to do distant hybridization. He produced an hybrid between two species of *Dianthu.*

  Dianthus caryophyllus Carnation × *D. barbatus* Sweet william
- ✰ Inter generic hybrid produced by Karpechenko, a Russian Scientist in 1928.

  Raphano brassica is the amphidiploid from a cross between Radish (*Raphanus sativus*) and cabbage (*Brassica oleraceae*), ultimately end up with useless combination.
- ✰ Triticale was produced by Rimpau in 1890 itself. Triticale is an amphidiploid obtained from cross between wheat and rye (man made cereal).
- ✰ Another example is *Saccharum* nobilisation involving three species.

Inter-specific Hybridization

Crossing between two species of the same genus.

Characteristics

- ✰ Efficient method of transferring desirable genes into cultivated plants from their related/wild species.
- ✰ Leads to introgression of important traits into cultivated varieties/species.

3 types of Crosses

1. **Fully Fertile:** between those species that have complete chromosomal homology. Chromosomes in such hybrids have normal pairing at meiosis.

   G. hirsutum (2n= 52) x *G. barbadense* (2n=52)

   G. arboreum (2n=26) x *G. herbaceum* (2n=26)

   T. aestivum (2n=42) x *T. compactum* (2n=42), club wheat.

   Avena sativa (2n=42) x *A. byzantiana* (2n=42), red oat.

   G. Max (2n=40) x *G. soja* (2n=40)

2. **Partially Fertile:** between those species which differ in chromosome number but have some chromosomes in common.

   T. aestivum (AA BB DD, 2n=42) x *T. durum* (AA BB, 2n=28)

   ↓

   F_1 AA BB D (14II + 7I)

   G. hirsutum (AA DD, 2n= 52) x *G. thurberi* (DD, 2n=26)

   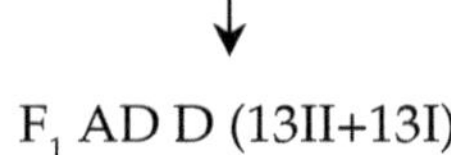

   F_1 AD D (13II+13I)

3. **Fully sterile:** between those species which don't have chromosomal homology. The lack of chromosome homology does not permit pairing between the chromosomes of two species during meiosis.

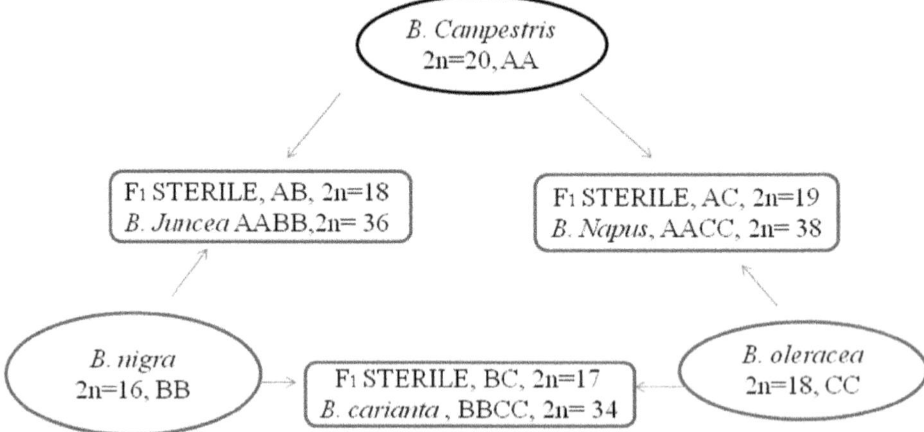

Figure 20: Fully Sterile Crosses: Bassica Triangle/Nagarharu U's Triangle.

- ✰ CLAUSEN and GOODSPEED- *N. sylvestris* (2n=24) x *N. tomentosa* (2n=24)
- ✰ *N. paniculata* (2n=24) x *N. undulata* (2n=24)
- ✰ HARLAND (1940)- *G. arboreum* (2n=26) x *G. thurberi* (2n=26)
- ✰ Brassica triangle/Nagarhelu U's triangle:

Barriers to Production of Distant Hybrids

In some cases, distant hybridisation may be obtained without an appreciable difficulty. *e.g. G. hirsutum* x *G. barbadense* crosses. But in majority of cases, F_1 hybrids may be obtained with variable degrees of difficulty and in many cases, hybrids may not be obtained with currently available technique.

e.g. Corchorus olitorius x *C. capsularis*

The difficulties encountered in the production of inter-specific hybrids may be grouped into three broad cases *viz.,*

1. **Failure of Zygote Formation/Cross Incompatibility/Pre-fertilization Barriers**: It may be due to failure of fertilization. It may be due to
   a) Slow pollen tube growth. *e.g.* Datura.
   b) Style may be longer than pollen tube. *e.g.* Maize and *Trypsaccum* crosses
   c) Pollen tubes of polyploid species may be thicker than diploid species and growth of pollen tube will be slower.
2. **Failure of Zygote Development/Hybrid Inviabilty:** In many cases fertilization takes place and zygote is produced. But the development of zygote is blocked at various stages. It may be due to
   a) **Lethal genes**: These lethal genes will act only in inter-specific crosses.

      e.g. Aegilops umbellulata has 3 lethal genes. These will act when crosses made between *A. umbellulata* and diploid wheat.
   b) **Genotypic disharmony between two parental genomes**: Genetic imbalance between two species may lead to death of the embryo.

      e.g. Crosses with *Gossypium davidsoni* show early embryonic mortality. This may be due to lethal gene action.
   c) **Chromosome elimination**: Chromosomes are gradually eliminated from the zygote. This has been due to mitotic irregularities. Elimination is reported in *Nicotiana, Hordeum, Triticum.*
   d) **Incompatibility Cytoplasm**: Cytoplasm of female parent may prevent embryo development.
   e) **Endosperm abortion**: The endosperm may abort during early embryo development.

      e.g. Triticum and *Secale* crosses; *Hordeum bulbosum* x *H.vulgare* crosses.
3. **Failure of hybrid seedling development/hybrid breakdown:** Due to chlorophyll deficiency seedlings may die. *e.g. Melilotus* hybrids.

Techniques for the Production of Distant Hybrids

1. First determine barrier to production of hybrid embryos.
2. In general the species with shorter style should be used as the female parent. Where ever this is not possible the style of the parent is to be cut off.
3. In some cases autopolyploidy may be helpful in inter-specific hybridization.

   e.g. Diploid *Brassicas* would not cross. Making them autotetraploids leads to easy crossing.
4. Wherever embryo abortion is there, embryo rescue technique can be adopted.
5. With species having different ploidy levels, are crossed, hybridization will be difficult.

   In such cases higher ploidy level species is used as female for crossing. Secondly, the chromosome number of wild species of F1 may be doubled to overcome sterility.

   e.g. Solanum tuberosum cross with wild species.
6. When two species cannot be crossed directly, a third species can be used as bridge species.

   Nicotiana repanda x N. sylvestris N. tobacum x N. sylvestris

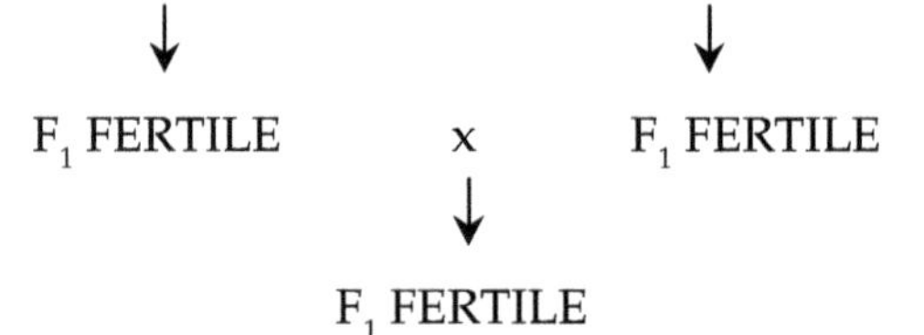

7. Use of growth regulators like IAA, 2, 4 - D at the flowering stage increases seed set.

Role of Wild Species in Crop Improvement

Many of our important crop species are allopolyploids. These crops evolved through distant hybridization. *e.g.* Groundnut, Ragi, Sugar cane, Cotton, Wheat, Brassicas, Tobacco, Potato. Many of the wild species are having resistance against both biotic and abiotic stresses which are successfully utilized for crop improvement.

Rice

Co 31 - *O. perennis* x GEB 24 Drought resistant.

IR 34 - a derivative of complex cross and *Oryza nivara* is one of the parents having resistance against grassy stunt virus. The following wild species are being used in breeding programme.

O. barthi	BLB resistance
O. longistaminata	Drought tolerance
O. rufipogan	Source of male sterility in rice.

Sorghum

Sorghum halapense 2n : 20 form crossed with CO.11 and the fodder sorghum variety Co27 was evolved.

S.nitidum : Highly resistant to shoot fly and having high dormancy. *S. stafii*: Having high dormancy occurs as weed in sorghum fields. *Sorghum sudanense* is used for evolving forage sorghum.

Bajra/Cumbu

Pennilsetum purpureum crossed with *P. glaucum* to evolve Cumbu Napier forage grass.

1. *P. glaucum* x *P. squamulatum*

   Forage grass combining frost resistance.
2. *P. glaucum* x *P. orientale*

   For the development of apomicts.
3. *P. glaucum* x *P. setaceum*

   Development of male sterile lines.

Sugarcane

Nobilisation of sugarcane involves the wild species *S.spontaneum*.

Red Gram

Cajanus cajan crossed with *C. lineata* and *C. scaraboides* to have resistance against wilt and also to induce male sterility.

Ground Nut

Arachis batizoccoi crossed with *A. hypogaea* to have rust resistant lines.

Arachis villosulicarpa - for increased number of pods.

A. monticola - for thin shelled condition

Sesamum

S. malabaricum is crossed with *S.indicum* to have male sterile lines as well as to have resistance against powdery mildew.

S. alatum: Resistance against powdery mildew and phyllody.

Cotton

By transferring *hirsutum* genome to the cytoplasm of wild species *G.harknessi* CGMS lines were obtained.

G. tomentosum: Resistant to drought, Jassids, lint fineness and strength.

G. barbadense var. *darwinii:* jassid (Tetraploid) resistant

G. hirsutum race *punctatum:* Resistant to black arm.

A number of diploid wild species are available having resistance against pest and diseases.

Potato

Cultivated tetraploid Potato *S.tuberosum* is obtained by natural crossing of diploid wild species *S.sparsipilum* with *S.vernii* followed by natural doubling. The following diploid species are used in breeding programme.

1. *S. ajanhuii:* Frost resistant
2. *S. phureja:* Non - dormant
3. *S. ptenomum:* Having 6 month dormancy, longer in duration.

Tobacco

Cultivated tobacco which is an amphidiploid was obtained by natural crossing with wild species and doubling.

Wild Species

N. debneyi – Resistant to root rot.

N. longiflora – Resistant to black shank disease.

N. glutinosa – Mosaic resistant for bridging two species.

N. digluta is used for bridging two species.

Application of Wide Hybridization in Crop Improvement

1. **Alien - Addition lines**: An alien addition line carries one chromosome pair from a different species in addition to normal diploid chromosome complement of the parent species. When *only one chromosome* from another species is present it is known as alien addition monosome. Alien addition lines have been produced in wheat, oats, tobacco and several other species. This is generally used to transfer disease resistance from wild species.
2. **Alien - Substitution lines**: An alien-substitution line has chromosome pair from a different species in the place of one chromosome pair of the recipient species. Alien substitution monosome has been developed in wheat, Cotton, tobacco. In tobacco, mosaic resistance gene N was transferred from *N. glutinosa* to *N. tabacum*.
3. **Transfer of small chromosome segments:** Transfer of small chromosome segments carrying specific desirable genes have been widely used in crop improvement programmes. *e.g.* Transfer of black arm resistance from *G. barbadense* to *G. hirsutum*.
4. **Quality Improvement:** By transferring genes from wild species quality has been improved. *e.g.* Genes for increased protein content in rice, soybean, oats and rye.

5. **Mode of reproduction**: In some cases incompatibility alleles from wild species can be transferred to cultivated species for hybrid seed production: *e.g.* Brassicas.
6. **Yield:** Increased yield through introgression of yield gene from a related wild species into cultivated species: *e.g.* Oats
7. **Transfer of cytoplasm**: Done by repeated back crossing. Mainly used for transferring male sterility into cultivated species. *e.g. Sesamum*
8. **Development of new crop species:** *e.g.Raphano brassica, Triticale.*
9. **Improvement in adaptation** ***etc.***

Limitations of WH

- ✰ Incompatible crosses.
- ✰ F_1 sterility.
- ✰ Problems in creating new sps.
- ✰ Undesirable linkage.
- ✰ Problem in transfer of recessive oligogenes and quantitative traits
- ✰ Lack of flowering in F_1.
- ✰ Dormancy in F_1 *etc.*

Pre-requisites for Hybridization

Breeder should have clear knowledge about the following before taking up hybridization.

1. Local conditions *i.e.* soil, climate, Agronomic practices and market requirements
2. Existing varieties of crops both local and introduced
3. Facilities like funds, land, labour and equipment
4. Plant material *i.e.* germplasm
5. Objectives: Well set objectives and planning

Hybridization Procedure or Steps involved in Hybridization

1. Objective

Based on the requirement, set your objective. Because based on the objective only the selection of parents is done. If it is resistance breeding one of the parents must be a donor.

2. Selection of Parents

Normal practice is, the female parent will be a locally adapted one in which we can bring in the plus genes. In case of inter-varietal hybridization geographically diverse parents will be selected so as to get superior segregants.

3. Evaluation of Parents

In case of parents which are new to the region they must be evaluated for their adaptability. Further to ensure homozygosity, they must be evaluated.

4. Sowing Plan

If the flowering duration is same, simultaneous sowing of both the parents can be done. Otherwise staggered sowing is to be followed. The normal practice is to raise the ovule parent in the centre of the plot in rows and on the border pollen parent for each combination.

5. Emasculation and Dusting

Emasculation is the removal of immature anthers from a bisexual flower. Depending on the crop the emasculation practice differs. Normal practice of hand emasculation and dusting of pollen is done. Depending on the time of anthesis the time of emasculation differs. For *e.g.* in rice the anthesis at Coimbatore takes place between 7.00 to 10.00 A.M. So the emasculation is done at around 6.30 A.M. and dusting of pollen is done immediately.

6. Labeling and Bagging

Immediately after hybridization put a label indicating the parents and date of crossing. Put appropriate cover to prevent foreign pollen, contamination.

7. Harvesting and Storage of Seeds

Normally 15-20 days after crossing the seeds will be set. In the case of pulses the crossed pods can be easily identified by the shrunken nature of pod and seed set will be reduced. Harvest of crossed seeds must be done on individual plant basis. Seeds collected from individual plants are to be stored in appropriate containers with proper label and stored.

From F_2 onwards the generations are known as segregating generations and they may be handled either by pedigree method of Bulk method or backcross method for evolving new varieties.

The advantages of growing hybrid:

- ☆ Hybrids are generally higher yielding than open-pollinated varieties, if grown under suitable conditions.
- ☆ Hybrids are uniform in color, maturity, and other plant characteristics, which enables farmer to carry out certain operations, such as harvesting at the same time.
- ☆ The uniformity of the grain harvested from hybrids can also have marketing advantages when sold to buyers with strict quality standards.

The disadvantages of growing hybrid:

- ☆ Hybrid seed is more expensive than open-pollinated seed.
- ☆ Farmers situated in a low potential environment and who cannot afford extra inputs such as fertilizer will not recover the costs of the hybrid seed.

- Fresh hybrid seed needs to be bought every planting season.
- The grain from a crop grown with hybrid seed should not be used for seed. The farmer cannot replant grain as seed without major reductions in yield, which might be a decrease of 30 per cent or more.
- The farmer might not always be able to source new seed in time for the planting season.

11

Methods of Breeding of Self-Pollinated Crops (Pedigree, Bulk, SSD, Back Cross, Multilines and Cultivar Blends)

Pedigree selection is a widely used method of breeding self-pollinated species (and even cross-pollinated species such as corn and other crops produced as hybrids). A key difference between pedigree selection and mass selection or pure-line selection is that hybridization is used to generate variability (for the base population), unlike the other methods in which production of genetic variation is not a feature. The method was first described by **H. Lowe** in 1927.

Pedigree Record and Maintenance

The pedigree may be defined as a description of the ancestors of an individual and it generally goes back to some distant ancestors. It is useful to know the relationship of two individuals and useful for selection of parents and prediction of outcome of the cross.

In the pedigree method, individual plants are selected from F_2 and subsequent generations, and their progenies are tested. During the entire operation a record of all parent offspring relationships is kept. This is known as pedigree record. Individual plant selection is continued till the progenies show no segregation. At this stage the selection is done among the progenies, multilocation tests are conducted and released as varieties.

For maintenance of pedigree record is the basic thing required is Crossing Ledger. This Ledger gives the details about parentage, Season in which the cross is made.

Example

Sl.No.	Cross Number	Parentage
1.	X S 1401	Co2 x MS 9804
2.	X S 1402	Co4 x C152
3.	X S 1403	Co1 x Co4

X = Cross

S = Summer Season

16 = Year (2016)

01 = Cross number

There are several ways to maintain the pedigree Record. The selection of plants starts from F_2 onwards. The details about selected plants can be recorded as follows. *e.g.* F_2 X S 1601 – 7, Progeny obtained from plant number 7 selected in F_2. F_3 X S 1601 - 7 – 4, Progeny obtained from plant number 4, selected from the F_3 progeny derived from the plant number 7 selected in F_2. This can be followed till F_4 or F_5 generations. After F_4 or F_5 the selected plants are bulked to form a family. In the pedigree record all the biometrical data like plant height, number of branches, No. of pods/plant, pod length, seeds/pod, pod weight, seed weight are recorded.

Applications

Pedigree selection is applicable to breeding species that allow individual plants to be observed, described, and harvested separately. It has been used to breed species including peanut, tobacco, tomato, and some cereals, especially where readily identifiable qualitative traits are targeted for improvement.

Procedure of Pedigree Method

1st Year

Cross is made between the parents possessing desirable characters.

2nd Year

Sow the F_1 seed giving wide spacing so that each F_1 plant produces more seeds. Raise as many F_1 plants as possible to produce large number of F_2 seeds. Harvest in bulk.

3rd Year

Grow 2000- 10000 plants of F_2 giving wide spacing for full expression of the characters, along with parents for comparison. Depending upon the facilities and objectives of the programme about 100-500 superior plants are selected. The values of selection depend on the skill of the breeder. He has to judge which F_2 plant will produce superior progeny for characters under consideration. The breeder develops this skill through close study of the crop for many generations. The selection in F_2 is done for simply inherited characters like head type, disease resistance *etc.*, and selection for characters governed by many genes like yield will be reserved for later generations. The selected plants are harvested separately and given serial numbers and description entered in pedigree registers.

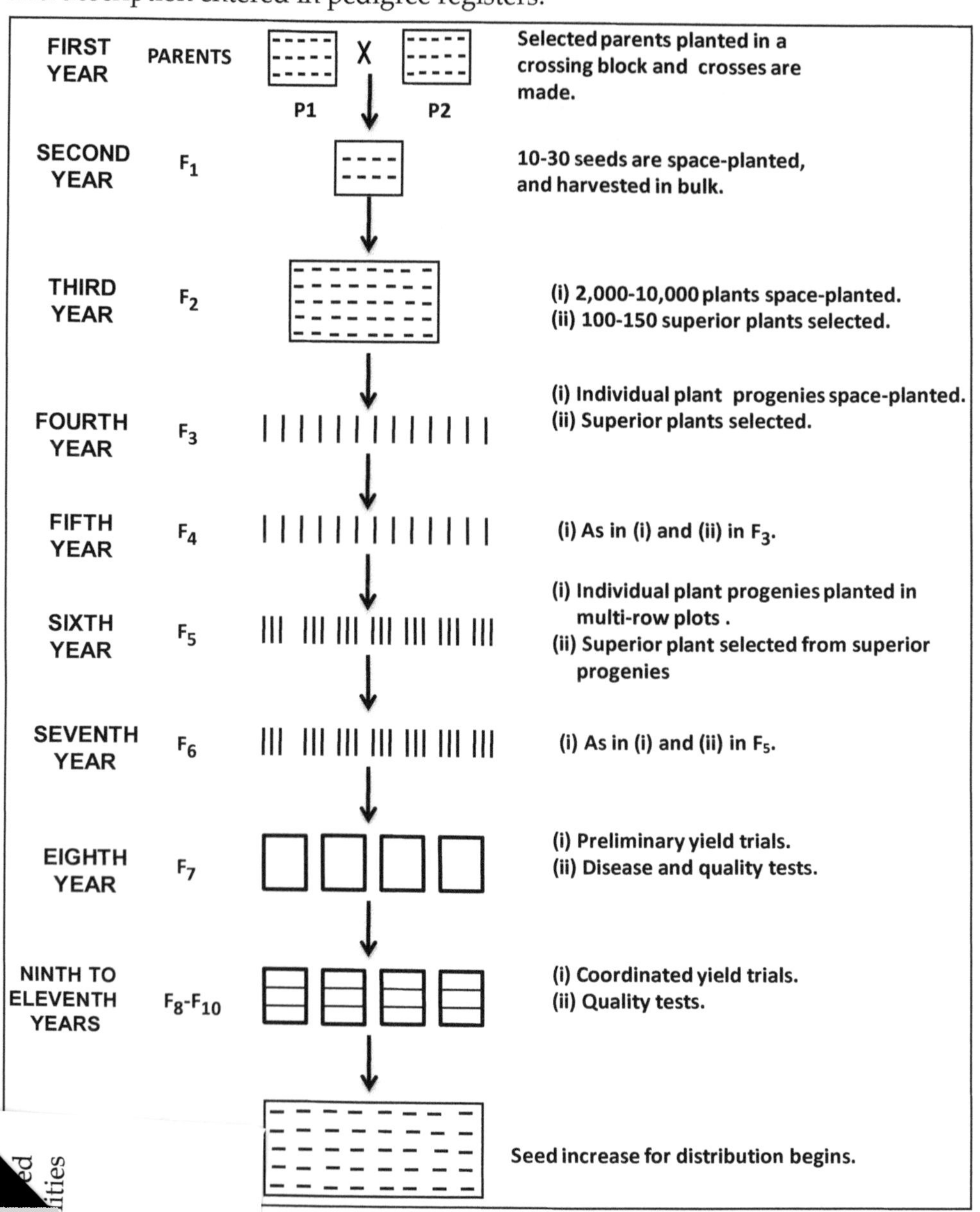

ified Schematic Representation of the Pedigree Method of
gating Generations from Crosses in Self-Pollinated Crops.

4th Year

Progeny rows of F_3 *i.e.* seeds of one selection plant in one row are space planted along with parents and checks. From superior progeny rows, individual plants with desirable characters are selected (about 50- 100 families and about 5 plants in each family and harvested separately). Diseased, lodging and undesirable progenies are discarded.

5th Year

F_4 plants raised again as head to row. Desirable plants are selected from desirable rows and harvested separately.

6th Year

F_5 plants raised in 3 row plots *i.e.* seeds of each selected plant sown in 3 rows. By this time many families might have become reasonably homozygous. For comparison check variety is grown for every 3 or 5 block. Progenies are evaluated for yield and the inferior ones are rejected. The number should be reduced to 25-50. Superior plants from superior progenies are selected. Plants from each progeny are bulked.

7th Tear

F_6 individual plant progenies are grown in multi-row plots and evaluated. Inferior progenies are rejected and superior progenies are selected. Plants of each progeny are harvested in bulk. Diseased and inferior plants from the progenies are removed.

8th Year

F_7 preliminary yield trial with 3 or more replications are conducted to identify superior lines. The progenies are evaluated for many characters including yield. Standard commercial varieties must be included as checks. Two to five outstanding lines are selected and advanced to coordinated yield trials.

9th, 10th and 11th Year

Selected lines are tested in several localities for 2 or 3 years for adaptation tests. Lines are evaluated for all characters mainly yield and disease resistance. A line that is superior to commercial variety in yield and other characters is selected.

11th and12th Year

Selected superior lines is named, multiplied and released as a new variety. Number of year can be reduced if generations are advanced during off seasons either in green house or under irrigated conditions.

Modifications of Pedigree Methods

Several modifications for the above described pedigree method are follow by breeders depending upon the crop, time and availability of funds and fa like labour, land *etc.*

Early Generation Tests

The objective of this test is to find out superior crosses and superior progenies in early generations *i.e.* in F_2 and F_3. We need not advance all the crossed and all selected progenies in each cross up to F_8. Much labour, time and cost would be saved by this early generation testing. A more reliable information about the potential crosses and progenies may be obtained by conducting replicated tests (preferably in more location) and evaluating them for yield and other characters in F_2 or F_3 itself. A desirable cross or progeny should have high mean yield, high genetic variance and high expected genetic advance under selection. Other crosses and progenies are rejected in the beginning *i.e.* F_2 and F_3 generations itself.

F_2 Progeny Testing

Another modification for pedigree method. In F_2 make as many single plants selections as possible. From F_3 to F_6 advance the progenies in bulk making selections of the progenies as a whole and discarding the inferior progenies. Thus each of the progeny is derived from the single plant selected in F_2 generation. In F_6 make single plant selections in each of the progeny. Compare the yields of the single plants with progenies from which they are selected. Select superior single plant progenies and advance to preliminary yield trials, multilocation tests *etc.*

There are two advantages

1. No. of crosses can be handled simultaneously
2. Natural selection operates from F_3 to F_6 since they are advanced in bulk.

Shuttle Breeding

This is followed especially in disease or insect resistance breeding. For *e.g.* at Coimbatore YMV in black gram is in epidemic form during *summe*r season only. Whereas at Vamban (Pudukkottai) the YMV is epidemic during *kharif* season. So instead of waiting for next summer at Coimbatore the materials can be tested at Vamban during *kharif* and thus one season is saved.

Mass Pedigree Method

This is another modified pedigree method. Crosses are made and further generations grown in bulk or as mass until suitable season occurs for making desirable selections against drought, insect and diseases *etc.* The population will be exposed to the natural conditions of selection. From the remaining population individual plants are selected and harvested progenies are evaluated for yield and other characters in preliminary yield trials and further generations are preceded as in pedigree method till release of variety. The advantages of both bulk and pedigree methods can be obtained and large number of crosses can be handled at a time. The disadvantage is that it takes a bit longer time.

Merits of Pedigree Method

1. It gives maximum opportunity for the breeder to use his skill and judgement in selection of plants.

2. It is well suited for the improvement of characters which can be easily identified and are simply inherited.
3. Transgressive segregation for yield and other quantitative characters may be recovered.
4. Information about the inheritance of characters and pedigree of lines can be obtained.
5. Inferior plants and progenies are eliminated in early generations.
6. It takes less time than bulk method to develop new variety.

Demerits of Pedigree Method

1. Valuable genotypes may be lost in early generations, if sufficient skill and knowledge are lacking in the breeder, at the time of selection.
2. No opportunity for natural selection.
3. Difficult to handle many crosses.
4. Maintenance of records, selections, growing progeny rows *etc.* are time consuming and laborious.

Achievements

Large number of varieties has been developed by pedigree method in many crops.

Wheat – NP -52, 120,125, 700 and 800 series

Rice – ADT – 25, Jaya, Padma, *etc.*

Cotton – Lakshmi, Digvijay, *etc.*

Sorghum – Co 18, RS 610 *etc.*,

Tobacco – NP 222

Sorghum – Co 18, Rs 610, *etc.*

Tobacco – NP 222

BULK BREEDING METHOD

Bulk population breeding is a strategy of crop improvement in which the natural selection effect is solicited more directly in the early generations of the procedure by delaying stringent artificial selection until later generations. The Swede, H. Nilsson-Ehle, developed the procedure. H. V. Harlan and colleagues provided an additional theoretical foundation for this method through their work in barley breeding in the 1940s. As proposed by Harlan and colleagues, the bulk method entails yield testing of the F_2 bulk progenies from crosses and discarding whole crosses based on yield performance. In other words, the primary objective is to stratify crosses for selection of parents based on yield values. The current application of the bulk method has a different objective.

Applications

It is a procedure used primarily for breeding self-pollinated species, but can be adapted to produce inbred populations for cross-pollinated species. It is most suitable for breeding species that are normally closely spaced in production (*e.g.*, small grains – wheat, barley). It is used for field bean and soybean. However, it is not suitable for improving fruit crops and many vegetables in which competitive ability is not desirable.

Procedure

The exact procedure for the bulk method would vary depending upon the objective of breeder. The following procedure is described for the isolation of homozygous lines. The breeder may introduce various modifications in the scheme to suit his needs.

Hybridization

Parents are selected according to the objective of the breeding programme. A simple or a complex cross is then made depending upon the number of parents involved.

F_1 Generation

F_1 is space- planted and harvested in bulk. The number of F_1 plants should be as large as possible; usually more than 20 plants should be grown.

F_2-F_6 Generations

F_2 to F_6 generations are planted at commercial seed rates and spacing's. These generations are harvested in bulk. During this period, environmental factors, disease and pest outbreaks would change the frequencies of different genotypes in the population. Artificial selection is generally not done. The population size should be as large as possible, preferably 30,000- 50,000 plants in each generation.

F_7 Generations

About 30-50 thousand plants are space- planted. 1000 to 5000 plants with superior phenotypes are selected and their seeds harvested separately. Selection is based on the phenotype of plants, grain characteristics, disease reaction, *etc.*

F_8 Generation

Individual plant progenies are grown in single or multi-row plots. Most of the progenies would be reasonably homozygous and are harvested in bulk. Weak and inferior progenies are rejected on the basis of visual evaluation. Only 100- 300 plant progenies with desirable characteristics are saved.

Some progenies which show segregation are generally rejected unless they are of great promise. In promising progenies, individual plants may be selected; preliminary yield trial will be delayed for one year in such cases.

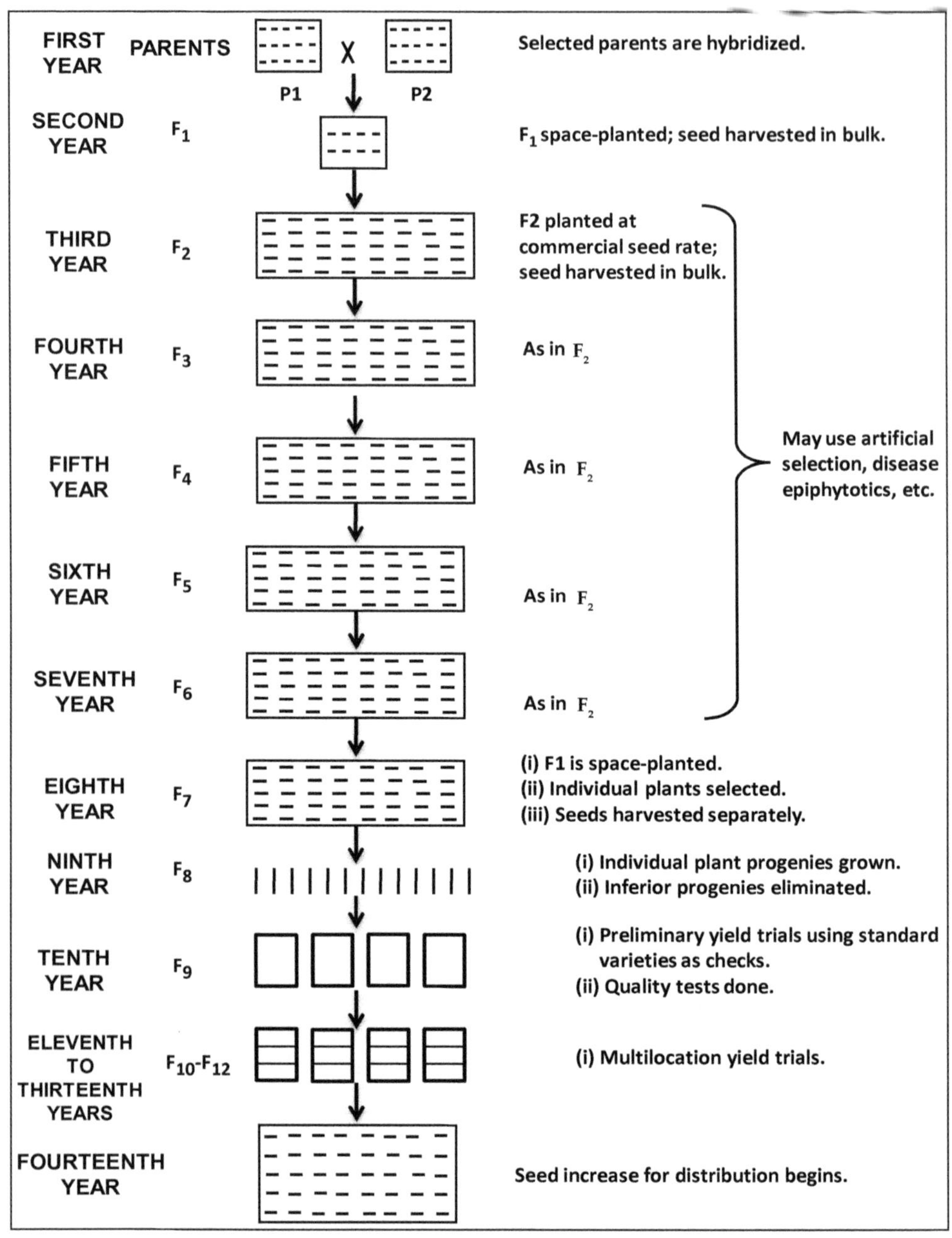

Figure 22: A Generalized Scheme of Bulk Method for the Isolation of Homozygous Lines of Self-Pollinated Crops.

F_9 Generation

Preliminary yield trial is conducted by using standard commercial varieties as checks. The progenies which are superior to the check are advanced. Quality test may be conducted to further reject undesirable progenies. The progenies are evaluated for height, lodging resistance, maturity date, disease resistance and other important characteristics of the crop species.

F_{10}-F_{13} Generations

Replicated yield trials are conducted over several locations using standard commercial varieties as checks. The lines are evaluated for important characteristics in addition to yield, disease resistance and quality. If a line is superior to the standard varieties in yield trials, it would be released as a new variety.

F_{14} Generation

Seed of the released variety is increased for distribution to the cultivators.

Modified Bulk Method

Here selection can be practiced in F_2 and F_3 and subsequent generations. There will not be any pedigree record but superior plants are selected bulked and carried forward. In F_4 superior plants are selected and harvested on single plant basis. In F_5 these single plants are studied in progeny rows and best progenies are selected and harvested. In F_6 PYT can be conducted to select best families. In subsequent generations regular trials can be conducted.

This modification of the bulk method provides an opportunity for the breeder to exercise his skill and judgement in selection. Further there is no maintenance of pedigree record which is another advantage.

Mass Pedigree Method

This was proposed by Harrington. It is a solution to one of the deficiencies in the pedigree method of breeding. For *e.g.* if the population is to be subjected to disease resistance screening like YMV and if there is no method to create artificial epiphytotic conditions, it is wasteful to study the population in pedigree method. Instead we can carry the population as a mass and test them when there is occurrence of the disease. When conditions are favourable for the disease, we can terminate the bulking and resort to single plant selection.

SINGLE-SEED DESCENT METHOD

The method of single-seed descent was born out of a need to speed up the breeding program by rapidly in breeding a population prior to beginning individual plant selection and evaluation, while reducing a loss of genotypes during the segregating generations. The concept was first proposed by *C. H. Goulden* in 1941 when he attained the F_6 generation in 2 years by reducing the number of generations grown from a plant to one or two, while conducting multiple plantings per year, using the greenhouse and off-season planting. H. W. Johnson and R. L. Bernard described the procedure of harvesting a single seed per plant for soybean in 1962. However, it was C. A. Brim who in 1966 provided a formal description of the procedure of single-seed descent, calling it a modified pedigree method.

The objective of single-seed- descent method is to rapidly advance the generations of crosses; at the end of the scheme, a random sample of homozygous or near homozygous genotypes/lines is obtained. F_2 and the subsequent generations are grown at very high plant densities as vigour of individual plants is not important. In each year, 2-3 generations may be raised using off- season nurseries and greenhouse facilities. The important features of this scheme are: (1) lack of selection, natural or artificial, till F_5 or F_6 till the population is reasonably homozygous, and (2) raising of F_3 and later generations from a bulk of one seed from each F_2 and the subsequent generation plant in order to ensure that each F_2 plant is represented in the population. As a result of the speed and economy, the single- seed- descent scheme is becoming increasingly popular with the breeders.

Merits

1. Advances the generation with the maximum possible speed in a conventional breeding method.
2. Requires very little space, effort and labour.
3. Makes the best use of greenhouse and off -season nursery facilities.
4. Ensures that the plants retained in the end population are random sample from the F_2 population.

Demerits

1. It does not permit any form of selection (which is implied in the scheme) during the segregating generations.
2. In each successive generation, the population size becomes progressively smaller due to poor germination and death of plants due to diseases, insect pests and accidents. In some crops, *e.g.*, pulses, plant loss may be one of the most serious problems of the scheme.

Applications

Growing plants in the greenhouse under artificial conditions tends to reduce flower size and increase cleistogamy. Consequently, single-seed descent is best for self-pollinated species. It is effective for breeding small grains as well as legumes, especially those that can tolerate close planting and still produce at least one seed per plant. Species that can be forced to mature rapidly are suitable for breeding by this method. It is widely used in soybean breeding to advance the early generation. One other major application of single-seed descent is in conjunction with other methods.

Procedure

Schematic representation of single-seed-descent method in showing in Figure 23.

Merits of Bulk Breeding Methods

1. The bulk method is simple, convenient and less expensive.
2. Since, each F_2 plant is equally represented till F_6, no chance of elimination of good genotypes in early generations.
3. Artificial or natural disease epiphytics, winter killing, high temperature

etc. eliminates undesirable types and increases the frequency of desirable type. Thus isolation of desirable types becomes easier.

4. Progenies select from long term bulks are superior to the selection from F_2 or short term bulk.
5. Since, little work and attention is needed in F_2 and subsequent generation more no. of crosses can be handled.
6. No pedigree records which saves time
7. Since large population are grown, transgressive segregants are more likely to appear and increase due to natural selection. Hence, there is a greater chance to isolate good segregants than pedigree method.

Demerits of Bulk Breeding Methods

1. The major disadvantage of bulk method is that it takes a much longer time to develop a new variety. Natural selection becomes important only after F_8 or F_{10}, and bulking may have to be done up to F_{20} or more. Thus the time required is considerably longer, and most breeders do not use the bulk method simply for this reason.
2. In short-term bulks, natural selection has little effect on the genetic composition of populations. But short -term bulks are useful for the isolation of homozygous lines and for specific objectives as in Harlan's mass - pedigree method.
3. It provides little opportunity for the breeder to exercise his skill or judgement in selection. But in the modified bulk method, the breeder has ample opportunity for practicing selection in the early segregating generations.
4. A large number of progenies have to be selected at the end of the bulking period.
5. Information on the inheritance of characters cannot be obtained which is often available from the pedigree method.
6. In some cases, at least, natural selection may act against the agronomically desirable types.

Achievements of Bulk Method

The method has been used to a limited extent is barley breeding in U.S.A. and more than 50 varieties were developed. They are: Arival, Beecher, Glacier, and Gem, originated from a cross: Atlas x Vaughn. The bulk was maintained for 7 to 8 generations.

☆ Narendra rai: *B. Juncea* variety.

BACK GROSS METHOD

Breeders of early 20th century engaged in the development of disease resistant varieties observed that pureline selections with genes for resistance from intra

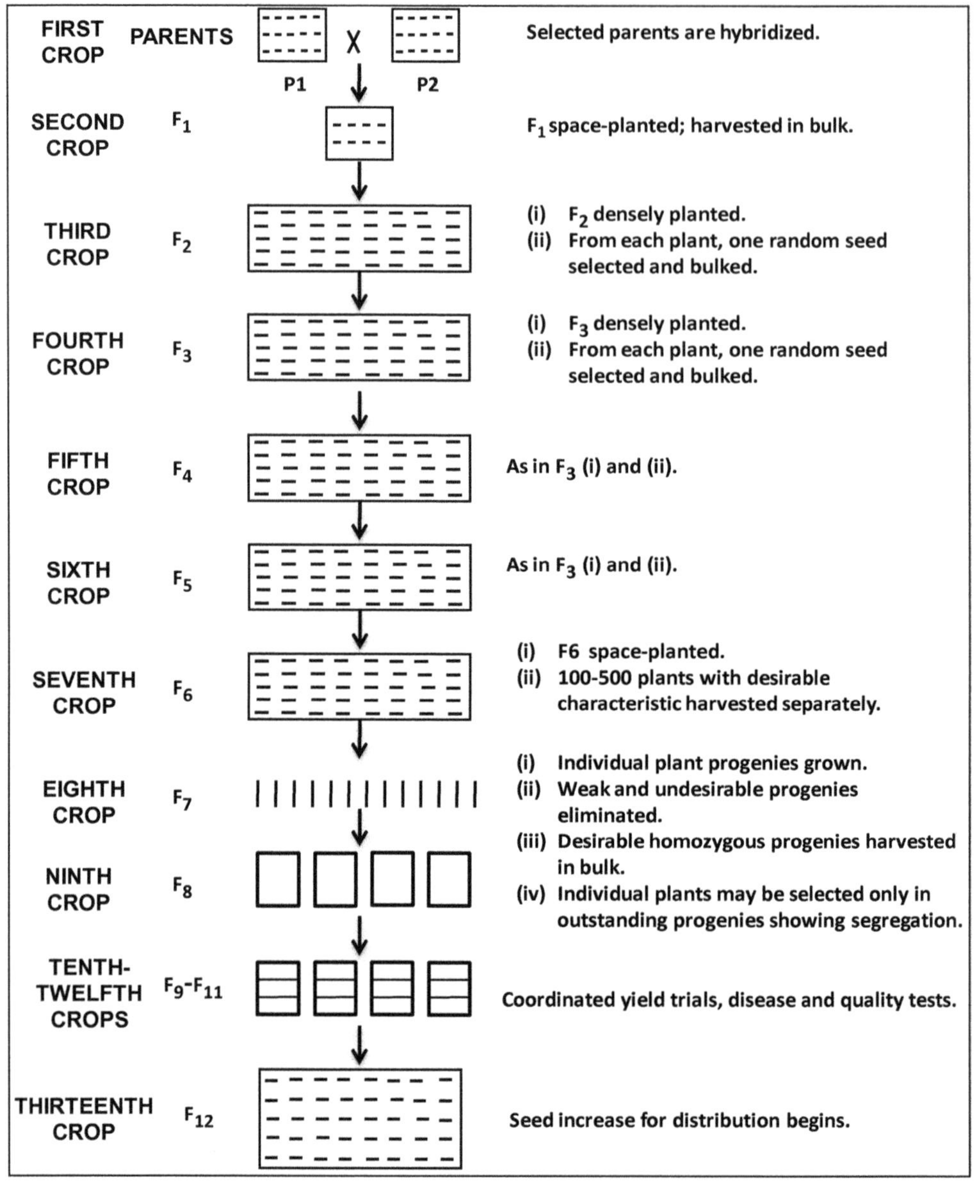

Figure 23: Schematic Representation of Single-Seed Descent Method.

-or inter- specific hybridization were inferior to the generally acceptance superior parent in yield or quality characteristic. To overcome this problem, (Harlan and Pope (1922) suggested the back cross method by which an undesirable allele at a particular locus is replaced by the desirable allele in otherwise elite variety. In other words, Back cross procedure conserves all good characteristics of a popular adapted variety and incorporates a desirable character from another variety.

Terminologies

Back Cross: A cross between a hybrid (F_1 or a segregating generation) and one of its parents is known as backcross.

Back Cross Method: In the back cross method, the hybrid and the progenies in the subsequent generations are repeatedly back crossed to one of their parents.

Objective: To improve or correct one or two specific defects of a high yielding variety, which is well adapted to the area and has other desirable characteristics.

Recipient Parent: Well adapted, high yielding variety, lacking one or two characters and hence receives these genes from other variety.

Donor Parent: The variety which donates one or two useful genes.

Recurrent Parent: Since the recipient parent is repeatedly used in the backcross programme, it is also known as the recurrent parent.

Non-recurrent Parent: The donor parent, on the other hand, is known as the non- recurrent parent because it is used only once in the breeding programme (for producing the F_1 hybrid).

Requirements of a Back Cross Program

1. Existence of a good recurrent parent variety which requires improvement in some qualitatively inherited character or a quantitative character with high heritability.
2. A suitable donor parent must be available possessing the character or characters to be transferred in a highly intense form.
3. High expressivity of the character under transfer through several back crosses in the genetic back ground of the recurrent parent.
4. The character to be transferred must have high heritability and preferably determined by one or few genes.
5. Simple testing technique for detecting the presence of the character under transfer.
6. Recovery of the recurrent genotype in a reasonable number of back cross generations.

Applications of Back Cross Method

1. **Inter varietal transfer of simply inherited characters**: characters governed by one or two major genes: *e.g.* Disease resistance, seed colour *etc.*

   Linkage drag: Failure of transfer of simply inherited characters like disease resistance by back cross method due to a tight linkage between the gene being transferred and some other undesirable gene.

2. **Inter varietal transfer of Quantitative characters:** Quantitative characters with high heritability can be transferred.

   e.g. Earliness, Plant height seed size, seed shape *etc.*

Table 32: Comparison between Bulk and Pedigree Methods

Sl.No.	*Pedigree Method*	*Bulk Method*
1.	Individual plants are selected in F_2 and the subsequent generations and individual plant progenies are grown.	F_2 and the subsequent generations are maintained as bulks.
2.	Artificial selection, artificial disease epidemics etc., are an integral part of the method.	Artificial selection, artificial disease epiphytotics etc., may be used to assist natural selection. In certain cases, artificial selection may be essential
3.	Natural selection does not play any role in the method.	Natural selection determines the composition of the populations at the end of the bulking period.
4.	Pedigree records have to be maintained which is often time consuming and laborious	No pedigree record is maintained.
5.	It generally takes 14-15 years to develop a new variety and to release it for cultivation.	It takes much longer for the development and release of a variety. The bulk population has to be maintained for more than 10 years for natural selection to act.
6.	Most widely used breeding method.	Used only to a limited extent.
7.	It demands close attention from the breeder from F_2 onwards as individual plant selections have to be made and pedigree records have to be maintained.	It is simple, convenient and inexpensive and does not require much attention from the breeder during the period of bulking.
8.	The segregating generations are space - planted to permit individual plant selection.	The bulk populations are generally planted at commercial planting rates.
9.	The size of population is usually smaller than that in the case of bulk method.	Large populations are grown. This and natural selection are expected to increase the chances of the recovery of transgressive segregants.

3. **Inter specific transfer of simply inherited characters:** Mostly disease resistance from related species into a cultivated species. *e.g.*
   a) Leaf and stem rust resistance from *Triticum timopheevii, T. monococcum, Aegilops speltoides* and *Rye* (*S. cereale*) to *T. aestivum*
   b) Black arm resistance from several *Gossypium* species to *G. hirsutum*
4. **Transfer of cytoplasm**: Back Cross method is used to transfer cytoplasm from one variety or species to another. This is especial desirable in cases of Cytoplasmic or Cytoplasmic-Genetic male sterility.

   e.g. Transfer of *T. timopheevii* cytoplasm to *T. aestivum.*
5. **Transresive segregation:** Back cross method may be modified to obtain transgressive segregants. It may be modified is one of the following two ways.
   a) The F_1 may be back crossed only 1 or at most 2 times to the recurrent parent leaving much heterozygosity for transgressive segregants to appear.
   b) Two or more recurrent parents may be used in the back cross programme to accumulate genes from them in to the back cross progeny. Such a modification of the back cross would produce a new variety that would not be exactly like any one of the recurrent parents.
6. **Production of Isogenic lines**: Isogenic lines are identical in their genotype, except for one gene. Such lines are useful in studying the effects of individual genes on yield and other characteristics. Isogenic lines and easily produced using the back cross method.
7. **Germplasm conversion:** Conversion of photosensitive germplasm lines (using as recurrent parent) to photo-insensitive line (using a photo insensitive line as a donor or non- recurrent parent).

Transfer of a Dominant Gene

Let us suppose that a high yielding and widely adapted variety A is susceptible to stem rust. Another variety B is resistant to stem rust, and that resistance to stem rust is dominant to susceptibility. A generalized scheme of the backcross programme for the transfer of rust resistance from variety B to variety A is given below:

Hybridization

Variety A is crossed to variety B. Generally, variety A should be used as the female parent. This would facilitate the identification of selfed plants, if any.

F_1 Generation

F_1 plants are backcrossed to variety A. Since all the F_1 plants will be heterozygous for rust resistance, selection for rust resistance is not necessary.

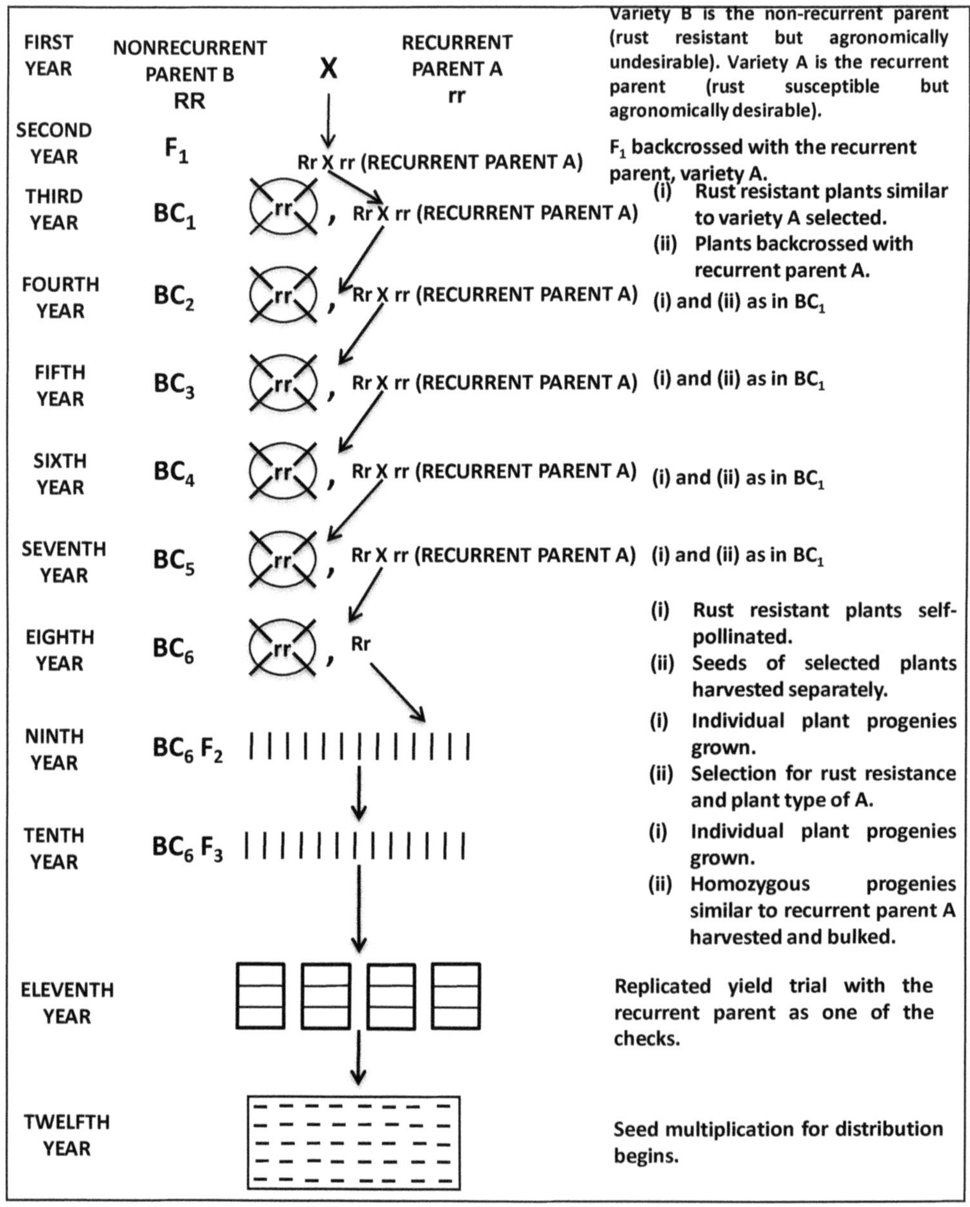

Figure 24: A Generalized Scheme for the Transfer of Domain Gene for Disease (Rust) Resistance through the Backcross Method in a Self-Pollinated Species.

(rr) Denotes that plants with rr genotypes are rejected.

First Backcross Generation (BC_1)

Half of the plants would be resistant and the remaining half would be susceptible to stem rust. Rust resistant plants are selected and backcrossed to variety A. BC_1 plants resistant to rust may be selected for their resemblance to variety A as well.

BC_2-BC_5 Generations

In each backcross generation, segregation would occur for rust resistance. Rust resistant plants are selected and backcrossed to the recurrent parent A. Selection for the plant type of variety A may be practiced, particularly in BC_2 and BC_3.

BC_6 Generation

On an average, the plants will have 98.4 per cent genes from variety A. Rust resistant plants are selected and selfed; their seeds are harvested separately.

BC_6 F_2 Generation

Individual plant progenies are grown. Progenies homozygous for rust resistance and similar to the plant type of variety A are harvested in bulk. Several similar progenies are mixed to constitute the new variety.

Yield Tests

The new variety is tested in a replicated yield trial along with the variety A as a check. Plant type, date of flowering, date of maturity, quality *etc.*, are critically evaluated. Ordinarily, the new variety would be identical to the variety A in performance. Detailed yield tests are, therefore, generally not required and the variety may directly be released for cultivation.

Transfer of a Recessive Gene

When rust resistance is due to a recessive gene, all the backcrosses cannot be made one after the other. After the first backcross, and after every two backcrosses, F_2 must be grown to identify rust resistant plants. The F_1 and the backcross progenies are not inoculated with rust because they would be susceptible to rust. Only the F_2 is tested for rust resistance. A generalized scheme for the transfer of a recessive gene for rust resistance is given below:

Hybridization

The recurrent parent is crossed with the rust resistant donor parent. The recurrent parent is generally used as the female parent.

F_1 Generation

F_1 plants are backcrossed to the recurrent parent.

BC_1 Generation

Since rust resistance is recessive, all the plants will be rust susceptible. Therefore, there is no test for rust resistance. All the plants are self-pollinated.

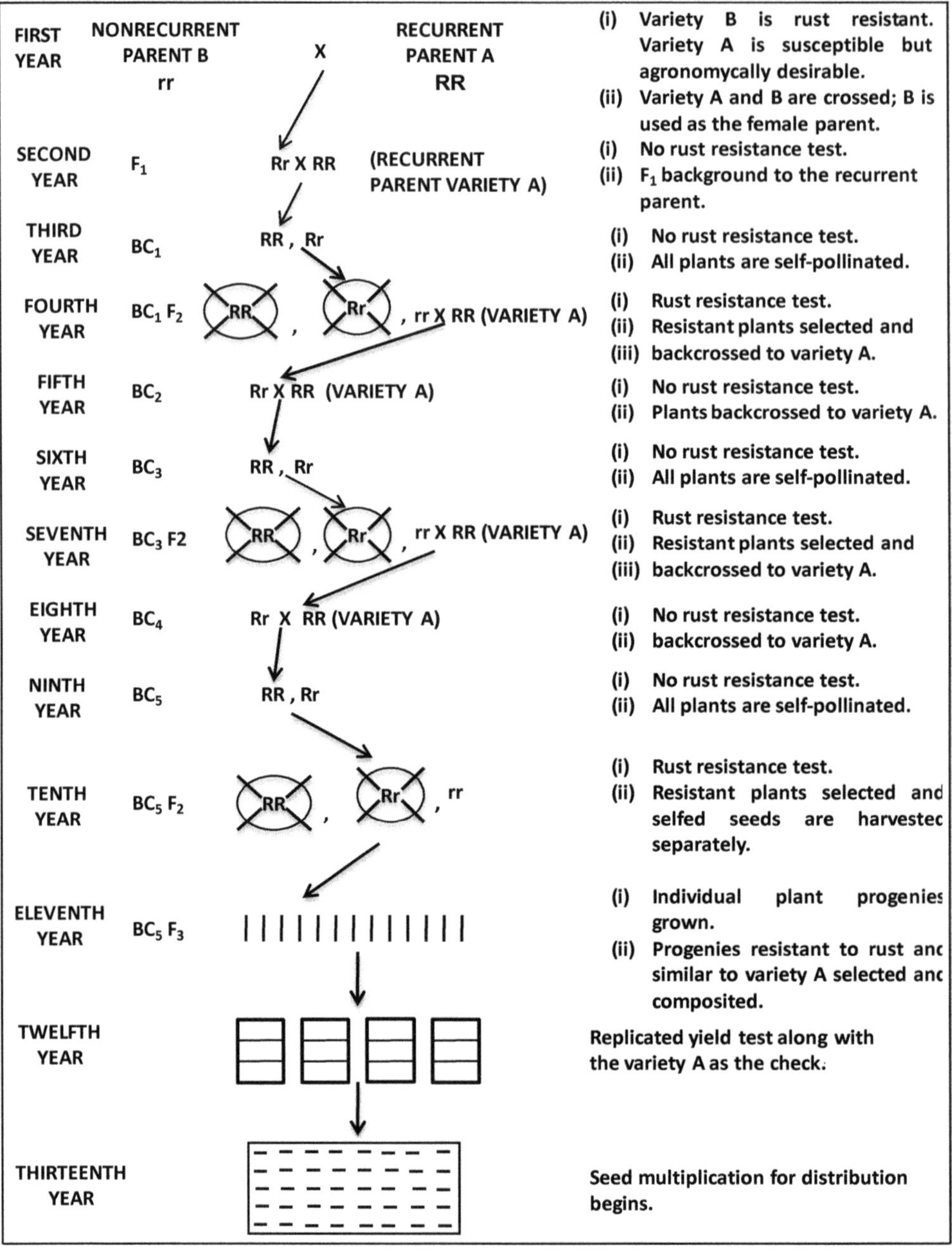

Figure 25: A Generalized Scheme for the Transfer of Recessive Gene for Disease Resistance (in this case rust resisance) through the Backcross Method in a Self-Pollinated Crop.

BC_1F_2 Generation

Plants are inoculated with rust spores. Rust resistant plants are selected and backcrossed with the recurrent parent. Selection is done for the plant type and other characteristics of the variety A.

BC_2 Generation

There is no rust resistance test. Plants are selected for their resemblance to the recurrent parent A, and backcrossed with the recurrent parent.

BC_3 Generation

There is no disease test. The plants are self-pollinated to raise F_2. Selection is usually done for the plant type of variety A.

BC_3F_2 Generation

Plants are inoculated with stem rust. Rust resistant plants resembling variety A are selected and backcrossed to variety A. Selection for plant type of A is generally effective.

BC_4 Generation

There is no rust resistance test. Plants are back-crossed to variety A.

BC_5 Generation

There is no rust test. Plants are self -pollinated to raise F_2 generation.

BC_5F_2 Generation

Plants are subjected to rust epidemic. A rigid selection is done for rust resistance and for the characteristics of variety A. Selfed seeds from the selected plants are harvested separately.

BC_5F_3 Generation

Individual plant progenies are grown and subjected to rust epiphytotic. A rigid selection is done for resistance to stem rust and for the characteristics of variety A. Seeds from several similar rust resistant homogeneous progenies are mixed to constitute the new variety.

Yield Tests

It is the same as in the case of transfer of a dominant gene.

Transfer of Two or More Characters into a Single Recurrent Parent

When two of more characters are to be transferred into the same variety, one of the following three approaches may be used.

Simultaneous Transfer

Genes for the different characteristics may be transferred simultaneously in the same backcross programme. The characters to be transferred are bought together into the hybrid by successively crossing each of the non-current parents to the recurrent parent or the hybrid thus produced. But in such a case, a larger backcross population would be needed than in the case of transfer of a single character. Further, the breeding programme may be delayed because the conditions necessary for the selection of all the characters may not occur every year. Sometimes, the two genes under transfer may be linked. In such a case, the transfer becomes very easy, and selection for only one gene may be necessary. Some examples of such a favourable linkage are; between the genes *Lr 24* and *Sr 24, Lr 19* and *Sr 25,* and *Lr 26* and *Sr 31*.

Stepwise Transfer

The recurrent parent is first improved for one character. The improved recurrent parent is then used as the recurrent parent in a backcross programme for the transfer of the second character. If additional characters are to be transferred, they are transferred one at a time in a stepwise fashion. This approach takes much longer time for the transfer of two or more characters.

Simultaneous but Separate Transfers

Each character is transferred to the same recurrent parent in simultaneous but separate backcross programmes. The resulting improved versions from the different programmes are then crossed together. Homozygous lines for the characters being transferred are then selected from the segregating generations using the pedigree method. This approach appears to be the most suitable of the three strategies.

Merits

1. The genotype of new variety is nearly identical with that of the recurrent parent, except for the genes transferred. Thus the outcome of a backcross programme is known beforehand and it can be reproduced any time in the future.
2. It is not necessary to test the variety developed by the backcross method in extensive yield tests because the performance of the recurrent parent is already known. This may save up to 5 years' time and a considerable expense.
3. The backcross programme is not dependent upon the environment, except for that needed for the selection of the character under transfer. Therefore, off-season nurseries and green-houses can be used to grow 2-3 generations each year. This would drastically reduce the time required for developing the new variety.
4. Much smaller populations are needed in the backcross method than in the case of pedigree method.
5. Defects, such as susceptibility to disease, of a well-adapted variety can be removed without affecting its performance and adaptability. Such a

Table 33: Comparison between Backcross and Pedigree Methods

Sl.No.	Pedigree Method	Backcross Method
1.	F_1 and the subsequent generations are allowed to self - pollinate	F_1 and the subsequent generations are backcrossed to the recurrent parent.
2.	The new variety developed by this method is different from the parents in agronomic and other characteristics.	Nearly identical to recurrent parent, except for the genes transferred.
3.	The new variety has to be extensively tested before release	Usually extensive testing is not necessary before release.
4.	The method aims at improving the yielding ability and other characteristics of the variety.	The method aims at improving specific defects of a well adapted, popular variety.
5.	It is useful in improving both qualitative and quantitative characters.	Useful for the transfer of both quantitative and qualitative characters provided they have high heritability.
6.	It is not suitable for gene transfer from related species and for producing substitution or addition lines.	It is the only useful method for gene transfers from related species and for producing addition and substitution lines.
7.	Hybridization is limited to the production of the F_1 generation.	Hybridization with the recurrent parent is necessary for producing every backcross generation.
8.	The F2 and the subsequent generations are much larger than those in the backcross method.	The backcross generation are small and usually consist of 20-100 plants in each generation.
9.	The procedure is the same for both dominant and recessive genes.	The procedures for the transfer of dominant and recessive genes are different.

variety is often preferred by the farmers and the industries to an entirely new variety because they know the recurrent variety well.

6. This is the only method for inter-specific gene transfer, and for the transfer of cytoplasm.
7. It may be modified so that transgressive segregation may occur for quantitative characters.

Demerits

1. The new variety generally cannot be superior to the recurrent parent, except for the character that is transferred.
2. Undesirable genes closely linked with the gene being transferred may also be transmitted to the new variety.
3. Hybridization has to be done for each backcross. This is often difficult, time taking and costly.
4. By the time the backcross programme improves it, the recurrent parent may have been replaced by other varieties superior in yielding ability and other characteristics.

Achievements

1. Kalyana Sona susceptible to leaf rust. Resistant has been transferred from several diverse sources *i.e.,* Robin, K1, Blue bird, Tobari, Frecor and HS-19.
2. Tift 23A is susceptible to downy mildew. The line backcrossed with MS-521A, MS- 541 A, MS- 570A resistant hybrids were produced.
3. Two cotton varieties *viz.,* 170-Co- 2 and 134-Co 2 were developed.

MULTILINE BREEDING AND CULTIVAR BLENDS

N. F. Jensen is credited with first using this breeding method in oat breeding in 1952 to achieve a more lasting form of disease resistance. Multilines are generally more expensive to produce than developing a synthetic cultivar, because each component line must be developed by a separate backcross.

Generally, pureline varieties are highly adapted to a limited area, but poorly adapted to wider regions. Further, their performance is not stable from year to year because of changes in weather and other environmental factors. Purelines often have only one or a few major genes for disease resistance, such as, rust resistance, which make them resistant to some races of the pathogen. New races are continuously produced in many pathogens which may overcome the resistance present in the pureline varieties. For example, Kalyan Sona wheat (*T. aestivum*) originally resistant to brown rust (leaf rust), soon became susceptible to new races of the pathogen.

To overcome these limitations, particularly the breakdown of resistance to disease, it was suggested to develop multiline varieties. *Multiline varieties are mixtures of several purelines of similar height, flowering and maturity dates, seed colour and agronomic characteristics, but having different genes for disease resistance.* The purelines constituting a multiline variety must be compatible, *i.e.*, they should not reduce the yielding ability of each other when grown in mixture.

In 1954, Borlaug suggested that several purelines with different resistance genes should be developed through back cross programmes using one recurrent parent. This is done by transferring disease resistance genes from several donor parents carrying different resistant genes to a single recurrent parent. Each donor parent is used in a separate backcross programme so that each line has different resistant gene or genes. Five to ten of these lines may be mixed depending upon the races of the pathogen prevalent in the area. If a line or lines become susceptible, they would be replaced by resistant lines. New lines would be developed when new sources of resistance become available. The breeder should keep several resistant lines in store for future use in the replacement of susceptible lines of multiline varieties.

Characteristics of a good Multiline

1. Its genetic diversity for vertical resistance genes for the concerned disease.
2. The vertical resistance genes should be strong enough.
3. It should have normal resistance to other diseases.
4. Components of multiline should be uniform for agronomic and other features.
5. It should have yield advantage.

Procedure

The backcross is the breeding method for developing multilines. The agronomically superior line is the recurrent parent, while the source of disease resistance constitutes the donor parent. To develop multilines by isolines, the first step is to derive a series of backcross derived isolines or near-isogenic lines (since true isolines are illusive because of linkage between genes of interest and other genes influencing other traits). The results of the procedure are two cultivars that contrast only in a specific feature. For disease resistance, each isoline should contribute resistance to a different physiological race (or group of races) of the disease.

The component lines of multilines are screened for disease resistance at multilocations. The breeder then selects resistant lines that are phenotypically uniform for selected traits of importance to the crop cultivar. The selected components are also evaluated for performance (yield ability), quality, and competing ability. Mixtures are composited annually based on disease patterns. It is suggested that at least 60 per cent of the mixture comprises isolines resistant to the prevalent disease races at the time. The proportion of the component lines are determined by taking into account the seed analysis (germination percentage, viability).

Advantages

1. A multiline provides protection to a broad spectrum of races of a disease-producing pathogen.
2. The cultivar is phenotypically uniform.

3. Multilines provide greater yield stability.
4. A multiline can be readily modified (reconstituted) by replacing a component line that becomes susceptible to the pathogen, with a new disease-resistant line.

Disadvantages

1. It takes a long time to develop all the isolines to be used in a multiline, making it laborious and expensive to produce.
2. Multilines are most effective in areas where there is a specialized disease pathogen that causes frequent severe damage to plants.
3. Maintaining the isoline is labor intensive.

Achievements

Multiline variety appears to be a useful approach to control diseases like rusts where new races are continuously produced. In India, three multiline varieties have been released in wheat (*T. aestivum*). Kalyan Sona, one of the most popular varieties in the late sixties, was used as the recurrent parent to produce these varieties. Variety 'KSML 3' consists of 8 lines having rust resistance genes from Robin, Ghanate, K1, Rend, Gabato, Blue Brid, Tobari *etc.* Multiline 'MLKS 11' is also a mixture of 8 lines; the resistance is derived from E 6254, E 6056, E 5868, Frecor, HS 19, E 4894 *etc.* The third variety, KML 7406 has 9 lines.

Cultivar Blends

Cultivars can be created with different genetic background (instead of one genetic backgroundt). When different genetic lines (2 or more cultivars) are combined, the mixture is a composite caklled a variety blend. Blends are less uniform in appearance than a pureline cultivar and they provide buffering effect against genotype x environmental interaction.

12

Genetic Structure of Cross-pollinated Crops

(System of Mating)

Cross-pollinated crops are highly heterozygous due to the free inter-mating among their plants. They are often referred to as random mating populations because each individual of the population has equal opportunity of mating with any other individual of that population. Such a population is also known as Mendelian population or panmictic population. A Mendelian population may be thought of having a gene pool consisting of all the gametes produced by the population. Thus gene pool may be defined as the sum total of all the genes present in a population. A population, in this case, consists of all such individuals that share the same gene pool, *i.e.*, have an opportunity to inter-mate with each other and contribute to the next generation of the population. Each generation of a Mendelian population may be considered to arise from a random sample of gametes from the gene pool of previous generation. For this reason, it is not possible to follow the inheritance of a gene in a Mendelian population by using the techniques of classical genetics. To understand the genetic make-up of such populations a sophisticated field of study, population genetics, has been developed. We shall examine the elementary principles of population genetics in order to understand the genetic composition of random mating populations, *i.e.*, cross-pollinated crops.

Hardy-Weinberg Law

The Hardy-Weinberg law is the fundamental law of population genetics and provides the basis for studying Mendelian populations. This law was independently developed by Hardy (1908) in England and Weinberg (1909) in Germany. **The Hardy-Weinberg law states that the gene and genotype frequencies in a Mendelian population remain constant generation after generation if there is no selection, mutation, migration or random drift.** The frequencies of the three genotypes for a locus with two alleles, say *A* and *a*, would be p^2 = AA, 2pq = Aa, and q^2 = aa; where p represents the frequency of *A* and q represents the frequency of *a* allele in the population, and the sum of p and q is one, (*i.e.*, p + q = 1). Such a population would be at equilibrium since the genotypic frequencies would be stable, that is, would not change, from one generation to the next. This equilibrium is known as Hardy-Weinberg equilibrium. A population is said to be at equilibrium when frequencies of the three genotypes, AA, Aa and aa are p2, 2pq and q2, respectively. Whether a population is at equilibrium or not can be easily determined using a chi-square test.

Hardy-Weinberg law can be easily explained with the help of an example. Let us consider a single gene with two alleles, *A* and *a*, in a random mating population. There would be three genotypes, AA, Aa and aa, for this gene in the population. Suppose the population has N individuals of which D individuals are AA, H individuals are Aa and R individuals are aa so that D +H + R = N. The total number of alleles at this locus in the population would be 2N since each individual has two alleles at a single locus. The total number of *A* alleles would be 2D+H because AA individuals have two *A* alleles each, while each Aa individual has only one *A* allele. The ration (2D+H)/2N are, therefore, the frequency of *A* allele in the population, and is represented by p. Similarly, the ratio (2R + H)/2N is the frequency of allele *a*, and is written as q. Therefore,

$$p = (2D + H)/2N \text{ or}$$
$$= (D + \tfrac{1}{2} H)/N \text{ and}$$
$$q = (2R + H)/2N \text{ or}$$
$$= (R + \tfrac{1}{2} H)/N$$

Therefore, p + q = 1

and p = 1 – q, o r q = 1 – p

The value of p and q are known as gene frequencies. Gene frequency is the proportion of an allele, *A* or *a*, in a random mating population. In other words, the proportion of gametes carrying an allele, *A* or *a*, is known as gene frequency. The genotype frequency or zygotic frequency is the proportion of a genotype, AA, Aa or aa, in the population. Random mating or random union of the two types of gametes would produce the AA, Aa and aa genotypes in a ratio proportionate to the frequencies of the gametes that united to produce them.

Factors Affecting Equilibrium Frequencies

The equilibrium in random mating populations is disturbed by migrations, mutation, selection and random drift. These factors are also referred to as evolutionary forces since they bring about changes in gene frequencies, which is essential for evolution to proceed. Obviously, a population in which gene and genotype frequencies remain constant over generations cannot evolve any further, unless its gene and genotype frequencies are disturbed.

Migration

Migration is the movement of individuals into a population from a different population. Migration may introduce new alleles into the population or may change the frequencies of existing alleles. The amount of change in gene frequency q will primarily depend upon two factors; first, the ratio of migrant individuals to those of the original population and second, the magnitude of difference between the values of q in the population and in the migrants. In plant breeding programmes, migration is represented by inter-varietal crosses, poly crosses, *etc.*, wherein the breeder brings together into a single population two or more separate populations.

Mutation

Mutation is a sudden heritable change in an organism and is generally due to a structural change in a gene. It is the ultimate source of all the variation present in biological materials. Mutation may produce a new allele not present in the population or may change the frequencies of existing alleles. However, since the mutation rate is generally very low, *i.e.*, approximately 10-6, the effects of mutation on gene frequency would be detectable only after a large number of generations. Therefore, in breeding populations such effects may be ignored.

Random Drift

Random drift or genetic drift is a random change in gene frequency due to sampling error. Random drift occurs in small populations because sampling error is greater in a smaller population than in a larger one. Ultimately, the frequency of one of the alleles becomes zero and that of the other allele becomes one. The allele with the frequency of one is said to be fixed in the population because there would be no further change in its frequency. It may be expected that in a small population all the genes would become homozygous, or would be fixed in due course of time. Breeding populations are generally small; hence a certain amount of genetic drift is bound to occur in them. The breeder cannot do anything to prevent this genetic drift, except to use very large populations, which is often not practicable. Alternatively, he may resort to phenotypic disassortative mating, which would again require time, labour and money.

Inbreeding

Mating between individuals sharing a common parent in their ancestry is known as inbreeding. In small populations, a certain amount of inbreeding is bound to occur. Inbreeding reduces the proportion of heterozygotes or heterozygosity and increases the frequency of homozygotes or homozygosity. The rate of decrease in heterozygosity is equal to ½ N (N = number of plant in the population) per generation in monoecious or hermaphrodite species. In dioecious species and in monoeicious species where self- pollination is prevented, the decrease in heterozygosity is somewhat lower; it is equal to ½(N=1) per generation. Thus in small populations, even with strict random mating or even with strict cross- pollination the frequency of homozygotes increases, while that of heterozygotes decreases due to inbreeding.

Selection

A differential reproduction rate of various genotypes is known as selection. In crop improvement, selection is very important because it allows the selected genotypes to reproduce, while the undesirable genotypes are eliminated. Thus the breeder is able to improve the various characteristics by selecting for the desirable types. In a random mating population, if plants with AA or aa genotypes are selected, the frequency of *A* allele in the selected population would be 1 or 0, respectively. It is assumed in this case that AA and aa genotypes would be identified without error. In the next generation, therefore, only *A* or *a* allele would be present, *i.e.*, the alleles would be fixed. Here selection against the remaining genotypes is complete, that is, these genotypes are not allowed to reproduce. In such cases, the disadvantage in reproduction, *i.e*, selection differential (S) is 1 and the fitness is zero for the remaining genotypes. The fitness of a genotype may be defined as its reproduction rate in relation to that of other genotypes. Generally, S has values less than one. Further, often it is not possible to identify the genotypes with certainty. The identification of genotypes is made difficult by dominance and due to less than 100 per cent heritability. This is particularly true for quantitative characters. As a result, selection is expected to change gene frequencies rather than to eliminate one or the other allele.

SYSTEMS OF MATING

To change the genetic composition of a population we have got different systems of mating.

1. Random Mating

Here the rate of reproduction of each individual is equal *i.e.* there is no selection and each male or female is equally likely to combine at random. This random mating is useful in plant breeding for the production and maintenance of synthetic and composite varieties, production of polycross varieties.

- ☆ Gene frequency remains constant.
- ☆ Variances of the character remain same.

2. Genetic Assortative Mating

Here the mating will be between individuals that are closely related by ancestry *i.e.*, mating between individuals having more or less similar genotype. It is otherwise known as inbreeding. The genetic assortative mating leads to:

- ☆ Increase in homozygosity and decrease in the hetretozygosity.
- ☆ Characters become fixed, except for the variation produced by the environment.
- ☆ Separation of population into lines.
- ☆ Genetic assortative mating is useful for the development of inbreds.

3. Genetic Disassortative Mating

- ✰ It is mating between individuals that are not closely related by ancestry. *e.g.* Inter-varietal and inter-specific crosses.
- ✰ Decrease in homozygosity and Increase in the hetretozygosity.

4. Phenotypic Assortative Mating

Mating between individuals which are phenotypically more similar. This type of mating leads to increase in homozygosity and division of population into two extremes *i.e.* there is highest and lowest phenotypes remain in the population and there is no intermediate types.

- ✰ This mating system is useful in the isolation of extreme phenotypes.
- ✰ It is used in some breeding schemes like recurrent selection.

5. Phenotypic Disassortative Mating

Mating between phenotypically dissimilar individuals. This system leads to maintenance of or increase in heterozygosity.

- ✰ There is some reduction in population variance, since it tends to produce intermediate phenotypes.
- ✰ Decrease the correlation between relatives due to increase in heterozygosity.

Selection in Cross-pollinated Crops

Selection in a random mating population is able to

1. Change the gene and genotypic frequency.
2. Production of new genotypes due to changed gene frequencies.
3. Cause a shift in the mean of population towards the direction of selection.
4. Change in the variance of population to some extent.

The magnitudes of these effects are influenced by the number of genes controlling the character, the degree of dominance, nature of gene action and to large extent heritability.

13

Breeding Methods for Cross-pollinated Crops (Population Improvement and Recurrent Selection)

Populations of cross-pollinated crops are highly heterozygous. When inbreeding is practiced they show severe inbreeding depression. So to avoid inbreeding depression and its undesirable effects, the breeding methods in the crop is designed in such a way that there will be a minimum inbreeding. The breeding methods commonly used in cross-pollinated crops may be broadly grouped into two categories.

I. Intra-population Improvement

A. Selection without Progeny Test

a) Mass Selection

b) Modified Mass Selection

- ☆ Detasseling
- ☆ Panmixis
- ☆ Stratified or Grid or Unit selection
- ☆ Honeycomb Design

B. Progeny Testing and Selection

a) **Half-sib Family Selection**

i) Ear to row

ii) Modified ear to row *etc.*

b) **Full-sib Family Selection**

c) **Inbred or Selfed Family Selection**

i) S_1 Self Family Selection

ii) S_2 Self Family Selection

II. Inter-population Improvement

1. Recurrent Selection

a) Simple Recurrent Selection

b) Reciprocal Recurrent Selection for GCA

c) Reciprocal Recurrent Selection SCA

d) Reciprocal Recurrent Selection.

2. Hybrids

3. Synthetics

4. Composites

Population Improvement

Cross-pollinated crops are highly heterozygous and heterogeneous. Consequently, they show varying degrees of inbreeding depression. Therefore, inbreeding should be avoided or kept to a minimum in cross-pollinated crops. Individual plants are heterozygous and their progeny would be heterogeneous and usually different from the parent, due to segregation and recombination. Therefore, desirable genes can be seldom fixed through selection in cross-pollinated crops except for highly heritable qualitative characters. Hence, the breeder aims of increasing the frequency of desirable alleles in the population. In cross-pollinated crops, the genotype of the individual plants is generally of little importance, especially in population improvement programmes but, the frequency of desirable alleles or genes in the population as a whole is more important.

Mass Selection

This is similar to the one which is practiced in self pollinated crops. A number of plants are selected based on their phenotype and open pollinated seed from them are bulked together to raise the next generation. The selection cycle is repeated one or more times to increase the frequency of favourable alleles. Such a selection is known as *Phenotypic Recurrent Selection*.

Merits

1. Simple and less time consuming.
2. Highly effective for character that are easily heritable -*e.g.* Plant height, duration *etc.*
3. It will have high adaptability because the base population is locally adapted one.

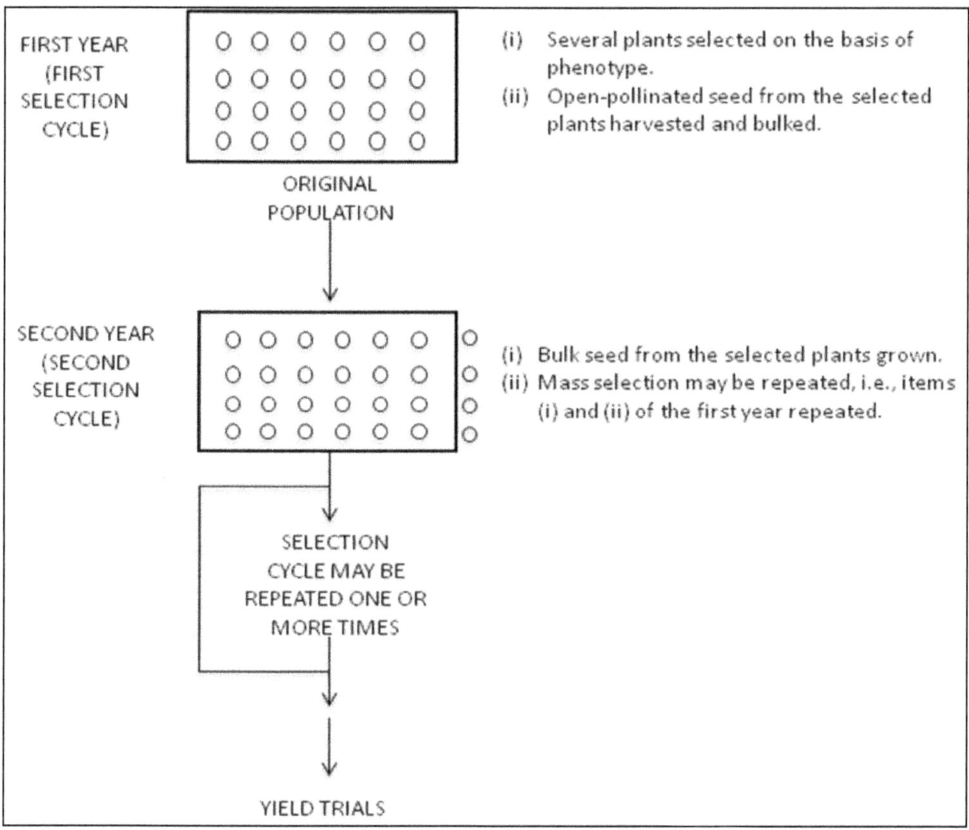

Figure 26: Mass selection as Applied to Cross-Pollinated Crops Species. When the selection is repeated one or more times, as oultlined here, the scheme is known as *Phenotypic* recurrent selection.

Demerits

1. Selection is based on phenotype only which is influenced by environment.
2. The selected plants are pollinated both by superior and inferior pollens present in the population.
3. High intensity of selection may lead reduction in population there by leading to inbreeding.

Modified Mass Selection

They proposed to overcome these defects.

1. **Detasseling**: This is practiced in maize. The inferior plants will be detasseled there by inferior pollen from base population is eliminated.
2. **Panmixis**: From the selected plants pollen will be collected and mixed together. This will be used to pollinate the selected plants. This ensures full control on pollen source.

3. **Stratified or grid system**: Proposed by C. O. Gardener, the field is divided into small grids (or subplots) with little environmental variance. An equal number of superior plants are selected from each grid for harvesting and bulking.
4. **Honeycomb design**: Proposed by A. Fasoulas, the planting pattern is triangular rather than the conventional rectangular pattern. Each single plant is at the centre of a regular hexagon, with six equidistant plants, and is compared to the other six equidistant plants.

The seeds from selected plants will be harvested and bulked to raise the next generation, by dividing the field into smaller plots, the environmental variation is minimized. This method is followed to improve maize crop.

Achievements

Mass and progeny selections have been extensively used for the improvement of cross-pollinated crops. The early varieties of bajra were developed through mass selection; some of the examples are; Bijapur 1, Jamnagar Giant, AF 3, S 530 and Pusa Moti; all these varieties were isolated from African introductions. Mass selection improved the yielding ability of toria by 30 per cent and oil content by 56 per cent; a further increase of 16 per cent in yield was obtained by using mass- pedigree method.

PROGENCY TESTING AND SELECTION

1) Half-sib Family Selection

Half-sibs are those which have one parent in common. Here only superior progenies are planted and allowed to open pollinate.

a) Ear-to-Row Method

This is the simplest scheme of half-sib selection applicable to cross-pollinated species. This is extensively used in maize. This method was developed by *Hopkins in 1908.*

Applications

Half-sib selection is widely used for breeding perennial forage grasses and legumes. A polycross mating system is used to generate the half-sib families from selected vegetatively maintained clones. The families are evaluated in replicated rows for 2–3 years. Selecting traits of high heritability (*e.g.*, oil and protein content of maize) is effective.

Ear- to-row Method of Selection is as follows:

1. A number of plants are selected on the basis of their phenotype. They are allowed to open pollinate and seeds are harvested on single plant basis.
2. A single row of 50 plants *i.e.* progeny row is raised from seeds harvested on single plant basis. The progeny rows are evaluated for desirable characters and superior progenies are identified.

3. Several phenotypically superior plants are selected from progeny rows. There is no control on pollination and plants are permitted to open pollinate.
4. Small progeny rows, as in item 2, are grown from the selected plants, and the process of selection is repeated.

Though this scheme in simple, there is no control over pollination of selected plants. Inferior pollen may pollinate the plants in the progeny row. To overcome this defect, the following method is suggested.

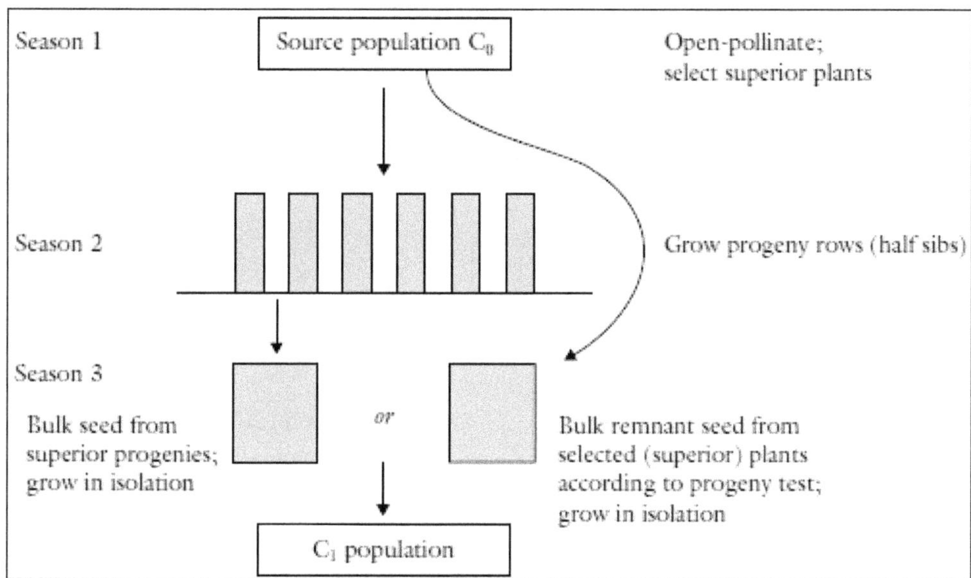

Figure 27: The Procedure of Ear-to-Row Method of Selection.

Modification of Ear-to Row Method

1. At the time of harvest of selected plants from base population on single plant basis, part of the seed is reserved.
2. While raising progeny rows, after reserving part of the seeds, the rest are sown in smaller progeny rows.
3. Study the performance of progenies in rows and identify the best ones.
4. After identifying the best progenies, the reserve seeds of the best progenies may be raised in progeny rows.
5. The progenies will be allowed for open pollination and best ones are selected.

2. Full-sib Family Selection

Full sibs are generated from biparental crosses using parents from the base population. The families are evaluated in a replicated trial to identify and select superior full-sib families, which are then recombined to initiate the next cycle.

Applications

Full-sib family selection has been used for maize improvement. A selection response per cycle of about 3.3 per cent has been recorded in maize.

Procedure

Cycle 0.

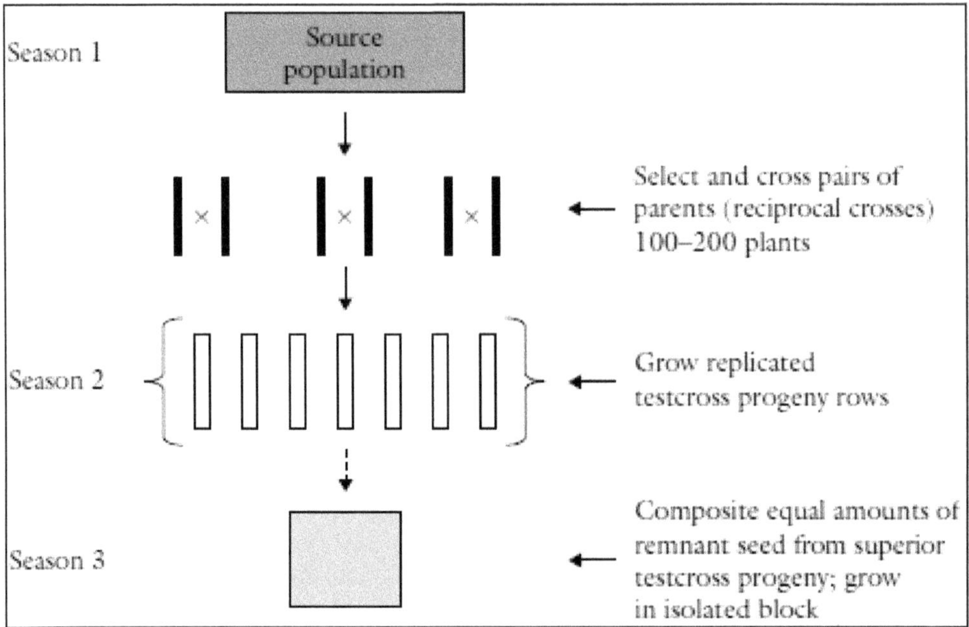

Figure 28: The Procedure of Full-sib Family Selection.

Season 1

Select random pairs of plants from the base population and inter-mate, pollinating one with the other (reciprocal pollination). Make between 100 and 200 biparental crosses. Save the remnant seed of each full-sib cross (*Figure 28*).

Season 2

Evaluate full-sib progenies in multiple location replicated trails. Select the promising half sibs (20–30).

Season 3

Recombine the selected full sibs.

3) Selfed (S_1 or S_2) Family Selection

An S_1 is a selfed plant from the base population. The key features are the generation of S_1 or S_2 families, evaluating them in replicated multienvironment trials, followed by recombination of remnant seed from selected families (*Figure 29*).

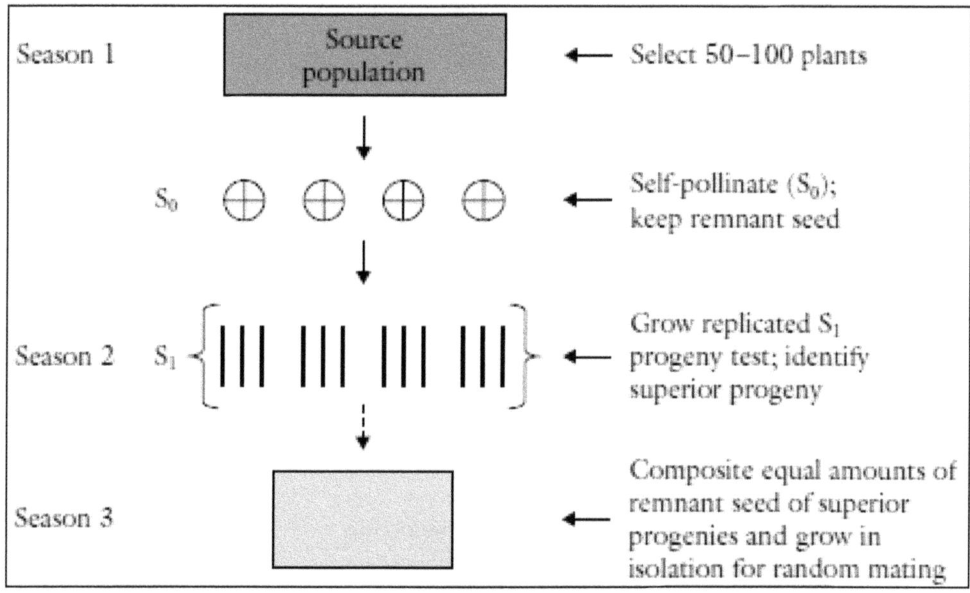

Figure 29: The Procedure of Selfed (S_1 or S_2) Family Selection.

Applications

The S_1 appears to be best suited for selfpollinated species (*e.g.*, wheat, soybean). It has been used in maize breeding. One cycle is completed in three seasons in S_1 and four seasons in S_2. A genetic gain per cycle of 3.3 per cent has been recorded.

Procedure

Season 1

Self-pollinate about 300 selected S0 plants. Harvest the selfed seed and keep the remnant seed of each S_1.

Season 2

Evaluate S_1 progeny rows to identify superior progenies.

Season 3

Random mate selected S_1 progenies to form a C_1 cycle population.

Merits of Progeny Testing and Selection

1. Selection based on progeny test and not on phenotype of individual plants.
2. In breeding can be avoided if care is taken raising a larger population for selection.
3. Selection scheme is simple.

Demerits

1. No control over pollen source. Selection is based only on maternal parent only.
2. Compared to mass selection, the cycle requires 2-3 years which is time consuming.

RECURRENT SELECTION

This is one of the breeding methods followed for the improvement of cross-pollinatedcrop. Here single plants are selected based on their phenotype or by progeny testing. The selected single plants are selfed. In the next generation they are inter-mated (cross in all possible combinations) to produce population for next cycle of selection.

History

Breeding schemes similar to recurrent selection were first suggested in 1919 by *Hayes and Garber* and independently by *East and Jones* in 1920. Critical data were not given by the above scientist. First detailed description of this type of breeding method was published by Jenkins in 1940 as a result of his experiments with early testing for GCA in maize. The method acquired its name in 1945 when, Hull suggested detailed scheme of Recurrent Selection for SCA. Hull (1952) defined recurrent selection as *"Method which involves reselection generation after generation with interbreeding of selects to provide for genetic recombination"*.

Applications

Recurrent selection may be used to establish a broad genetic base in a breeding program. Because of multiple opportunities for inter-mating, the breeder may add new germplasm during the procedure when the genetic base of the population rapidly narrows after selection cycles. Research has indicated that recurrent selection is superior to classic breeding when linkage disequilibrium exists. In fact, the procedure is even more effective when epistatic interactions enhance the selective advantage of new recombinants. Recurrent selection is applied to legumes (*e.g.*, groundnut, soybean) as well as grasses (*e.g.*, barley, oats).

Types of Recurrent Selection

There are four basic recurrent selection schemes, based on how plants with the desired traits are identified.

1. Simple Recurrent Selection

This is similar to mass selection with 1 or 2 years per cycle. The procedure does not involve the use of a tester. Selection is based on phenotypic scores. This procedure is also called *Phenotypic Recurrent Selection*.

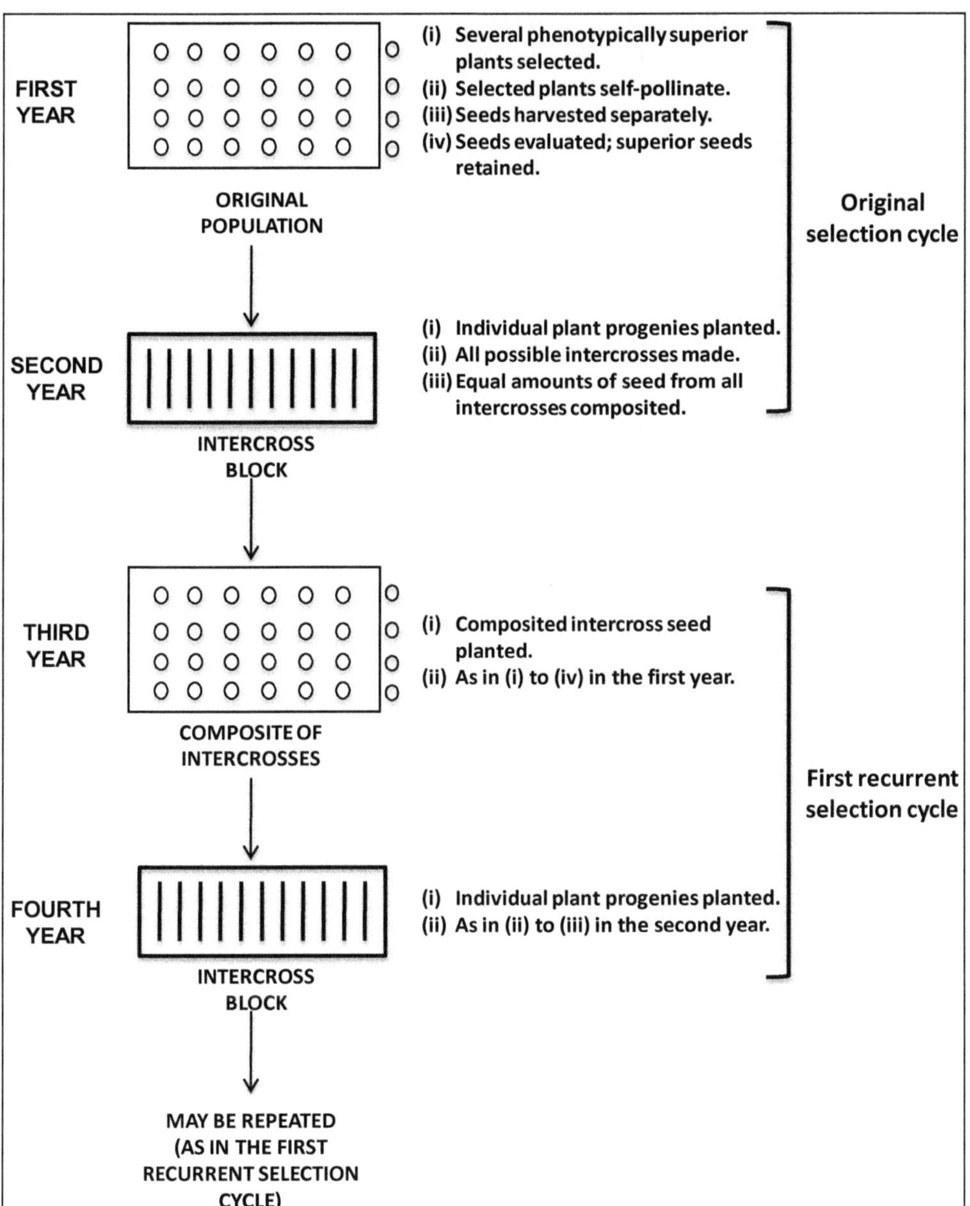

Figure 30: Simple Recurrent Selection when the Characters Under Selection can be Easily Evaluated on Basis of Phenotype of the Selected Plants for Form the Selfed Seed Obtained from them.

☆ **I year:** Several phenotypically superior plants are selected and selfed. Harvested separately and evaluated. Seed of superior plants retained and the rest are discarded.

- **II year:** Individual plant progeny rows are raised. The progeny rows are inter crossed in all possible combinations. Equal amounts of seed from each cross is taken and mixed. This forms the source for next selection cycle. (Original selection cycle).
- **III year:** Seed obtained in II year is planted. Superior plants selection, selfed and harvested separately. Superior plants retained and the rest discarded. (First selection cycle).
- **IV year:** Progeny rows are raised. Inter crossed in all possible ways. Equal amount of seed from each cross is composited. This mixed seed forms the source for next selection cycle.

The procedure ma y be repeated for another one or two selection cycles. The effectiveness of the simple recurrent selection was published with data by Sprague and Brimhall (1950). This is useful for characters that can be measured on individual plants and having high heritability. The procedure is to be modified suitably for characters which cannot be measured on individual plants.

2. Recurrent Selection for General Combining Ability

This is a half-sib progeny test procedure in which a wide genetic-based cultivar is used as a tester. The testcross performance is evaluated in replicated trials prior to selection.

Procedure

- **I year:** Several phenotypically superior plants are selected from source population. Each selected plant is selfed as well as crossed to a tester with broad genetic base. The selfed seeds are harvested separately and saved for planting in the third year. The test crossed seeds also harvested separately.
- **II year:** A replicated yield trial is conducted using the test crossed seeds. At the end the superior progenies are identified.
- **III year:** Selfed seed (from the first year) of the plants that produced superior progenies on the basis of yield trial of second year is planted in separate progeny rows. These progenies are inter crossed in all possible combinations. Equal amount of seed from each intercross is composited to raise the source population for next selection cycle.
- **IV year:** Source population is raised from the composited seeds. Several phenotypically superior plants are selected. They are selfed and crossed to a tester (broad genetic base) selfed seed harvested separately and saved for planted in third year and all test cross seeds also harvest separately.
- **V Year:** Repeat as in second year
- **VI year:** Repeat as in third year. This completes the first selection cycle. The second and third selection cycle may be initiated if necessary.

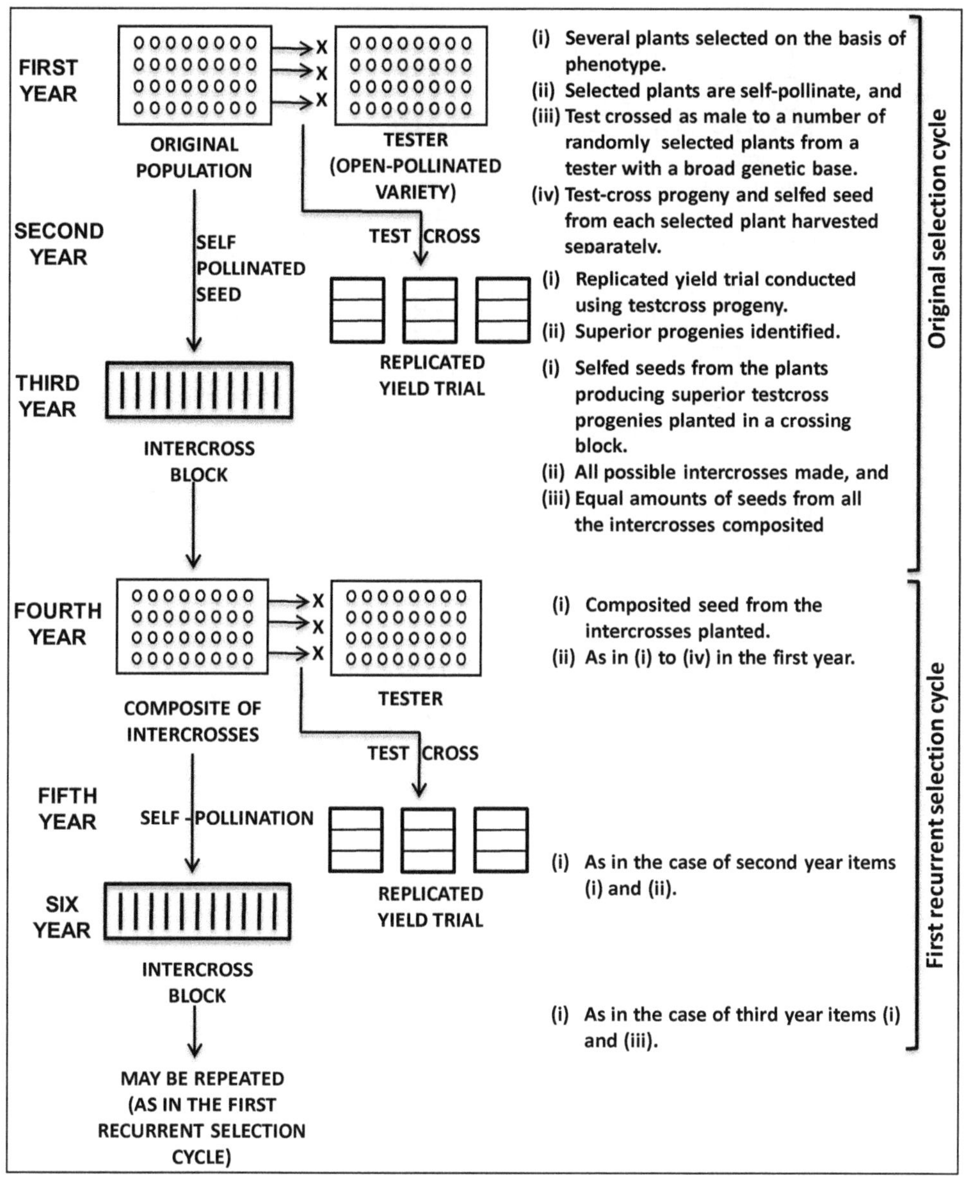

Figure 31: Recurrent Selection from General Combining Ability. In the case of recurrent selection for specific combining ability, an ibred is used as a tester in the place of an open-pollinated variety; the rest of the scheme remains the same.

The Recurrent Selection for GCA

- ✰ May be used for improving the yielding ability of the population and the end product may be released as a synthetic variety or,
- ✰ May be used for increasing the frequency of desirable genes in the population and the population may be used for isolating superior inbreds.

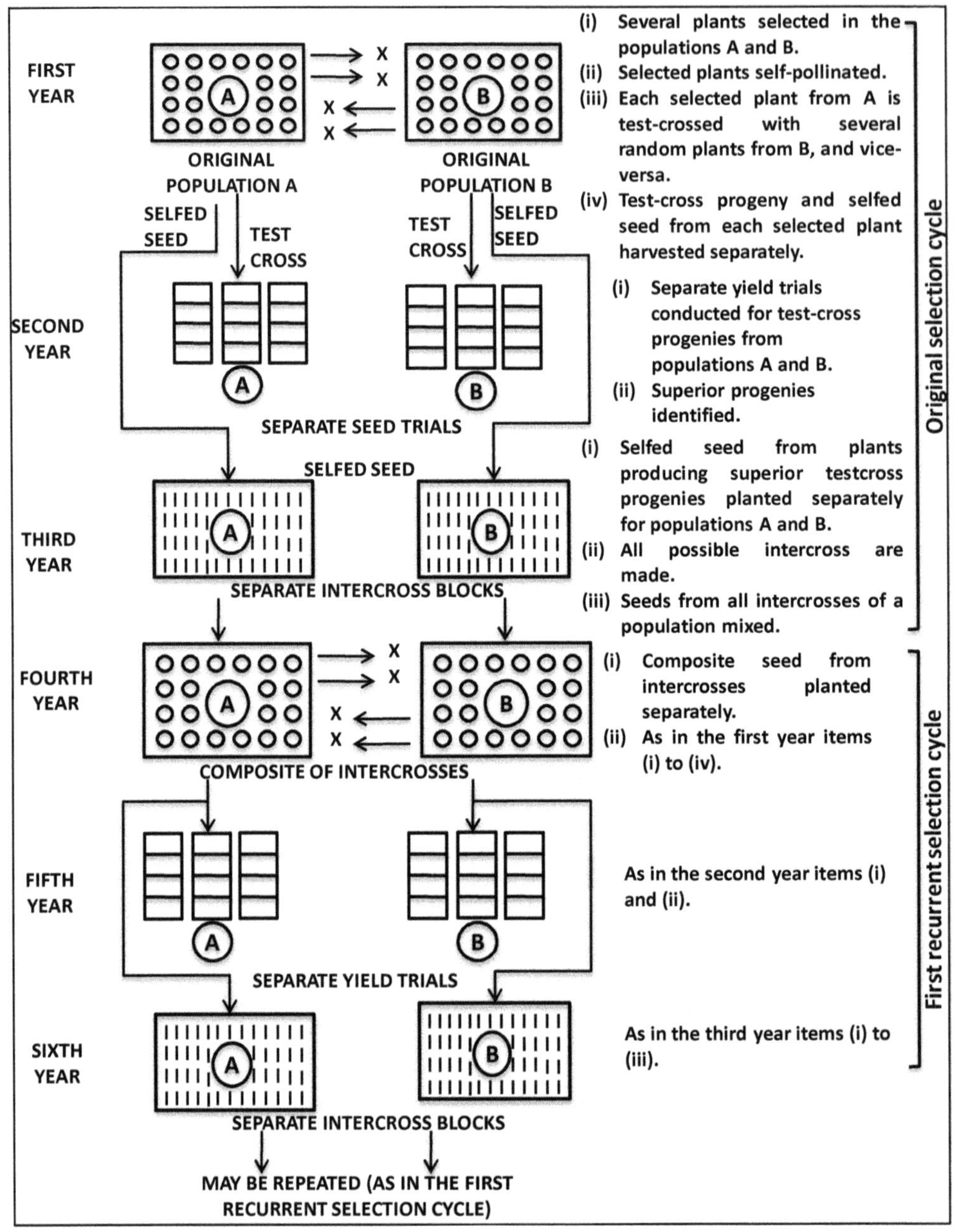

Figure 32: Reciprocal Recurrent Selection. Two populations A and B, with broad genetic bas serve as testers for each other.

3. Recurrent Selection for Specific Combining Ability

This scheme uses an inbred line (narrow genetic base) for a tester. The testcross performance is evaluated in replicated trails before selection. The recurrent selection for SCA was first proposed by Hull in 1945. The objective is the isolate from a population such lines that will combine well with a given inbred useful for selecting lines for SCA.

The procedure for recurrent selection for SCA is identical with that of recurrent selection for GCA, expect that the tester used here is an inbred (narrow genetic base).

4. Reciprocal Recurrent Selections

This scheme is capable of exploiting both general and specific combining ability. It entails two heterozygous populations, each serving as a tester for the other. Reciprocal recurrent selection was first proposed by Comstock, Robinson and Harvey in 1949. This would be useful.

1. For selecting both for SCA and GCA
2. For improving two source population simultaneously.

Procedure

- ☆ **I year:** Two source populations (A and B) are taken; several phenotypically superior plants are selected from each population. Each of the selected plant is selfed. Each of the selected plant from source A is crossed with random plants from B. Similar each of the selected plants from B are crossed with random plants of A. Plants of A will act as tester for B and vice-versa. The selfed seed is harvested separately and saved for planting in III year. Top crossed seed from each plant is also harvested separately.
- ☆ **II year:** Two replicated yield trials are conducted, progeny rows of test cross seeds of population A in one plot and test cross seeds of population B in another plot are raised. Plants (I year) producing superior progenies (in II year) are identified.
- ☆ **III year:** Selfed seed (saved in I year) from plants selected on the basis of evaluation of progeny rows (in the II year) is planted in plant to row progeny in two crossing plots. Seeds of selected plants from population A in one plot and that of the B in another plot. All possible intercrosses among the progeny rows in each plot are made. Equal amount of seed from all intercrosses from the crossing plot A is mixed to raise the source population of 'A' next year. Similarly equal amount of seed from inter crosses of plot B is mixed to raise source population B next year. This completes original selection cycle.
- ☆ **IV year:** Source populations of A and B are raised from composited seeds of A and B (III year). Operations of the first year *i.e.* selection of plants, selfing and crossing with the plants of other population *etc.* are done.
- ☆ **V year:** operations as in second year are repeated.
- ☆ **VI year:** Operations as in third year are repeated. This completes first selection cycle. The populations may be subjected to further selection cycles, if necessary by repeating the procedure outlined above.

Use of Reciprocal Recurrent Selection

1. Two populations are developed by this method
2. They may be inter-mated to produce a superior population with broad genetic base. This is similar to a varietal cross but in this case the populations have been subjected to selection for combining ability (GCA and SCA)
3. Inbreds may be developed from populations A and B. These inbreds may be crossed to produce a single cross or double cross hybrids.

Conclusion on the Efficiency of different Recurrent Selection Schemes

1. If dominance is incomplete Recurrent selection for GCA and reciprocal recurrent selection are equal but both are superior to recurrent selection for SCA.
2. If dominance is complete the three methods are equal
3. If over dominance is present Reciprocal recurrent selection and recurrent selection for SCA are equally effective both are superior to recurrent selection for GCA.

The Advantages of Recurrent Selections are

1. The rate of inbreeding can be kept at low level.
2. The frequency of favourable genes in the population will be increased and so.
3. The chance of obtaining satisfactory individuals from the population will be increased because greater opportunity for recombination is present.

Combining Ability

Ability of a strain to produce superior progeny when crossed with other trains.

General Combining Ability

- ☆ Average performance of a strain/genotype in a series of cross combinations.
- ☆ *e. g.*: A x B, A x C, A x D, A x E......, Average performance of strain A.
- ☆ It helps in selection of suitable parents for hybridization.
- ☆ The tester will have a broad genetic base.

Specific Combining Ability

- ☆ Performance of strain/genotype in a specific cross.
- ☆ *e.g.*: A X B
- ☆ It helps in identification of superior cross combinations.
- ☆ The testing will be on inbred.

14

Synthetics and Composites Varieties

Synthetic variety is produced by crossing in all combinations a number of lines that combine well with each other. Once synthesized, a synthetic is maintained by open- pollination in isolation.

Some breeders use the terms synthetic variety in a restricted sense: a synthetic variety is regularly reconstructed from the parental lines and is not maintained by open-pollination.

By definition, a synthetic variety consists of all possible crosses among a number of lines that combine well with each other. The lines that make up a synthetic variety may be inbred lines, clones, open- pollinated varieties, short - term inbred lines or other populations tested for GCA or for combining ability with each other.

Applications

The possibility of commercial utilization of synthetic varieties in maize was first suggested by Hayes and Garber in 1919. Synthetic varieties have been of great value in the breeding for those cross-pollinated crops where pollination control is difficult, *e.g.*, forage crop species, many clonal crops like cacao, alfalfa (*M. Sativa*), clovers (*Trifoulim* sp.) *etc.* The maize improvement programme in India now places a considerable emphasis on synthetic varieties. The maize programme of CIMMYT, Mexico, is based on population improvement; the end-product of such a programme is usually a synthetic variety.

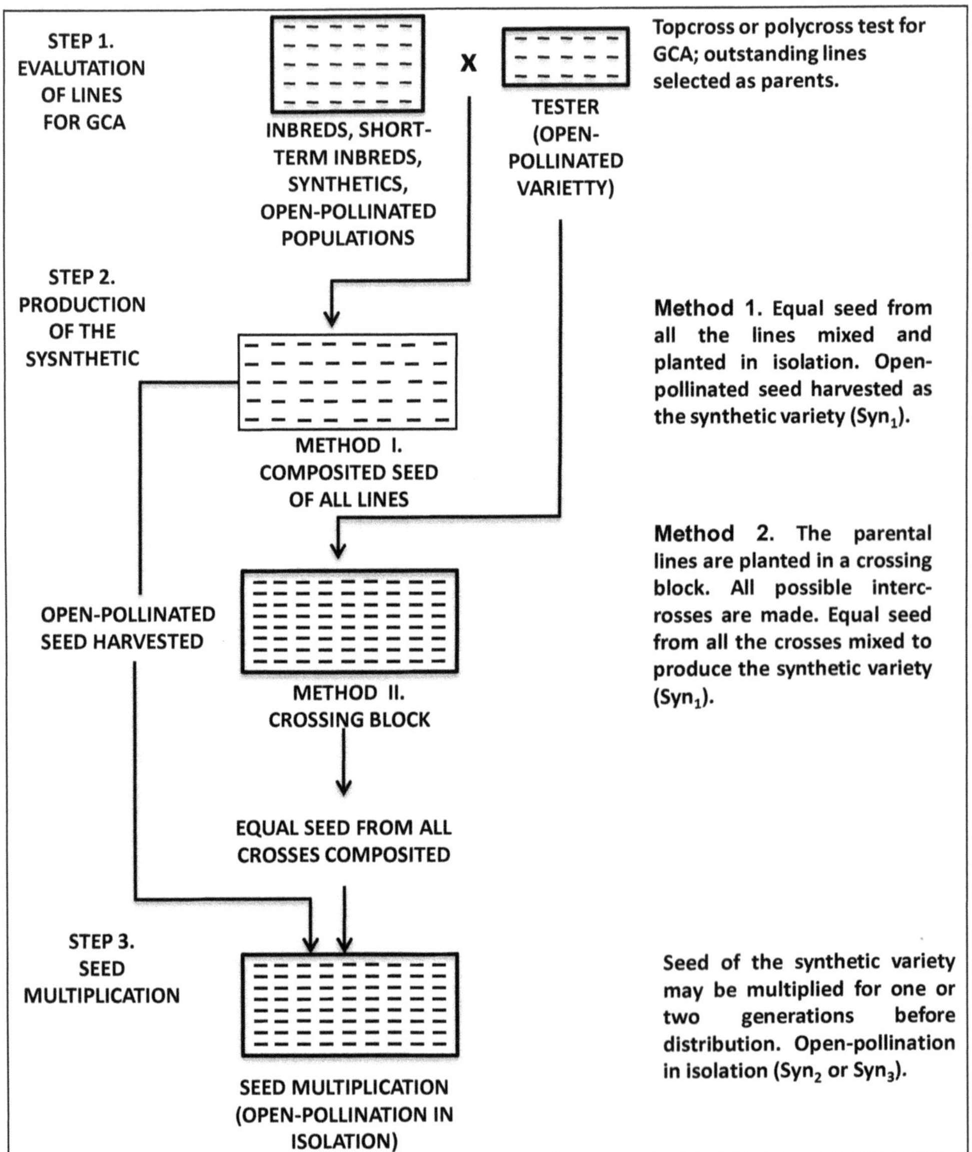

Figure 33: Steps Involved in the Production of Synthetic Varieties.

Steps Involved in Development of Synthetic Variety

Evaluation of Lines for GCA

GCA of the lines to be used as the parents of synthetic varieties is generally estimated by top cross or polycross test. The lines areevaluated for GCA because synthetic varieties exploit that portion of heterosis, which is produced by GCA. Polycross refers to the progeny of a line produced by pollination with random sample pollen from a number of selected lines. Polycross test is the most commonly used test in forage crops. Polycross progeny are generally produced by open-pollination in isolation among the selected lines. The lines that have high GCA are selected as parents of a synthetic variety.

Production of a Synthetic Variety

A synthetic variety may be produced in one of the following two ways.

1. Equal amounts of seeds from the parental lines are mixed and planted in isolation.

   Open- pollination is allowed and is expected to produce crosses in all combinations.

   The seed from this population is harvested in bulk; the population raised from this seed is the Syn1 generation.

2. All possible crosses among the selected lines are made in isolation. Equal amounts of seed from each cross are composited to produce the synthetic variety. The population derived from this composited seed is known as the Syn_1 generation.

Multiplication of Synthetic Varieties

After a synthetic variety has been synthesized, it is generally multiplied in isolation for one or more generations before its distribution for cultivation. This is done to produce commercial quantities of seed, and is a common practice in most of the crops, *e.g.*, grasses, clovers, maize *etc.* But in some crops, *e.g.*, sugar beets, the synthetic varieties are distributed without seed increase, *i.e.*, in the Syn_1 generation.

The open-pollinated progeny from the Syn_1 generation is termed as Syn_2, that from Syn_2 as Syn_3 *etc.* The performance of Syn_2 is expected to be lower than that of Syn_1 due to the production of new genotypes and a decrease in the level of heterozygosity as a consequence of random mating. However, there would not be a noticeable decline in the subsequent generations produced by open-pollination since the zygotic equilibrium for any gene is reached after one generation of random mating. The synthetic varieties are usually maintained by open- pollinated, and may be further improved through population improvement schemes, particularly through recurrent selection.

Factors Determining Performance of Synthetic Varieties

The yield of Syn_2 would be less than that of Syn_1 due to loss in heterozygocity as a result of random mating.

The decrease in yield ability of Syn_2 would depend on:

1. The number of inbred lines.
2. Mean performance of inbreds.
3. Mean performance of F_1 crosses.

Sewall Wright, 1922, suggested the formula for predicting the performance of Syn_2.

$Syn_2 = Syn_1 (Syn_1 - Syn_0)/n$

where,

n = number of parental lines

How to improve the performance of Syn_2 ?

There 3 ways:

1. By increasing the number of lines.
2. By increasing the performance of $Syn_{1.}$
3. By improving the performance of parental lines.

Merits of Synthetic Varieties

1. Synthetic varieties offer a feasible means of utilizing heterosis in crop species where pollination control is difficult. In such species, the production of hybrid varieties would not be commercially viable.
2. The farmer can use the grain produced from a synthetic variety as seed to raise the next crop.
3. In variable environments, synthetics are likely to do better than hybrid varieties. This expectation is based on the wider genetic base of synthetic varieties in comparison to that of hybrid varieties.
4. The cost of seed in the case of synthetic varieties is relatively lower than that of hybrid varieties.
5. Seed production of hybrid varieties is a more skilled operation than that of synthetic varieties.
6. Synthetic varieties are good reservoirs of genetic variability. The composites and germplasm complexes also serve as gene reservoirs.
7. There is good evidence that the performance of synthetic varieties can be considerably improved through population improvement without appreciably reducing variability.

Demerits of Synthetic Varieties

1. The performance of synthetic varieties is usually lower than that of the single or double cross hybrids. This is because synthetics exploit only GCA, while the hybrid varieties exploit both GCA and SCA.
2. The performance of synthetics is adversely affected by lines with relatively poorer GCA. Such lines often have to be included to increase the number of parental lines making up the synthetic as lines with outstanding GCA are limited in number.
3. Synthetics can be produced and maintained only in cross-pollinated crop species, while hybrid varieties can be produced both in self- and cross-pollinated crops.

Achievements

Synthetic varieties have been widely used in forage crops and in crops where pollination control is difficult. The maize breeding programme at CIMMYT, Mexico and the pearl millet programme at ICRISAT, Hyderabad are based on synthetic varieties generally developed through population improvement. The maize breeding programme in India is placing increasingly greater emphasis on the production of synthetic varieties.

COMPOSITE VARIETIES

In cross-pollinated crops, the mixture of genotypes from several sources that is maintained in bulk from the one generation to the next generation is referred as composite variety. Composite are constituted by seed mixture of advanced generation material of inter-varietal or inter racial crosses. The lines used to produce a composite variety are rarely tested for combining ability with each other. Consequently, the yields of composite varieties cannot be predicted in advance for the obvious reason that the yields of all the F_1's among the component lines are not available. Like synthetics, composites are commercial varieties and are maintained by open- pollination in isolation.

Procedure of Developing Composite Variety (Figure 34)

1. **1st year:** Selection of base material, selfing and inter-mating in all possible combinations and harvesting crossed seeds separately.
2. **2nd year:** Evaluation of F_1 crosses in replicated trial using hybrid or open pollinated variety as check.
3. **3rd to 5th year:** Evaluation of F_2, F_3 and F_4 generations in replicated trials using standard check, identification of crosses exhibiting little or no inbreeding depression.
4. **6th to 7th year:** Mixing parental seed of superior crosses in equal quantity to constitute composite variety, seed multiplication by open pollination in isolation.
5. **8th year:** Release of new composite variety, distribution of seed to the farmers for commercial cultivation.

Merits and Demerits

Same as in case of synthetic varieties.

Achievements

- ☆ In India, the first composite varieties were released in 1967; the six maize composites were, Ambar, Jawahar, Kisan, Vikaram, Sona and Vijay.
- ☆ Composite 1, has been evolved in *Brassica campestris* Var. *toria.* It was developed by compositing 10 elite *toria* strains; it matures in 100 days, exhibits profuse branching and yields about 11 q/ha seed, which contains about 40 per cent oil.

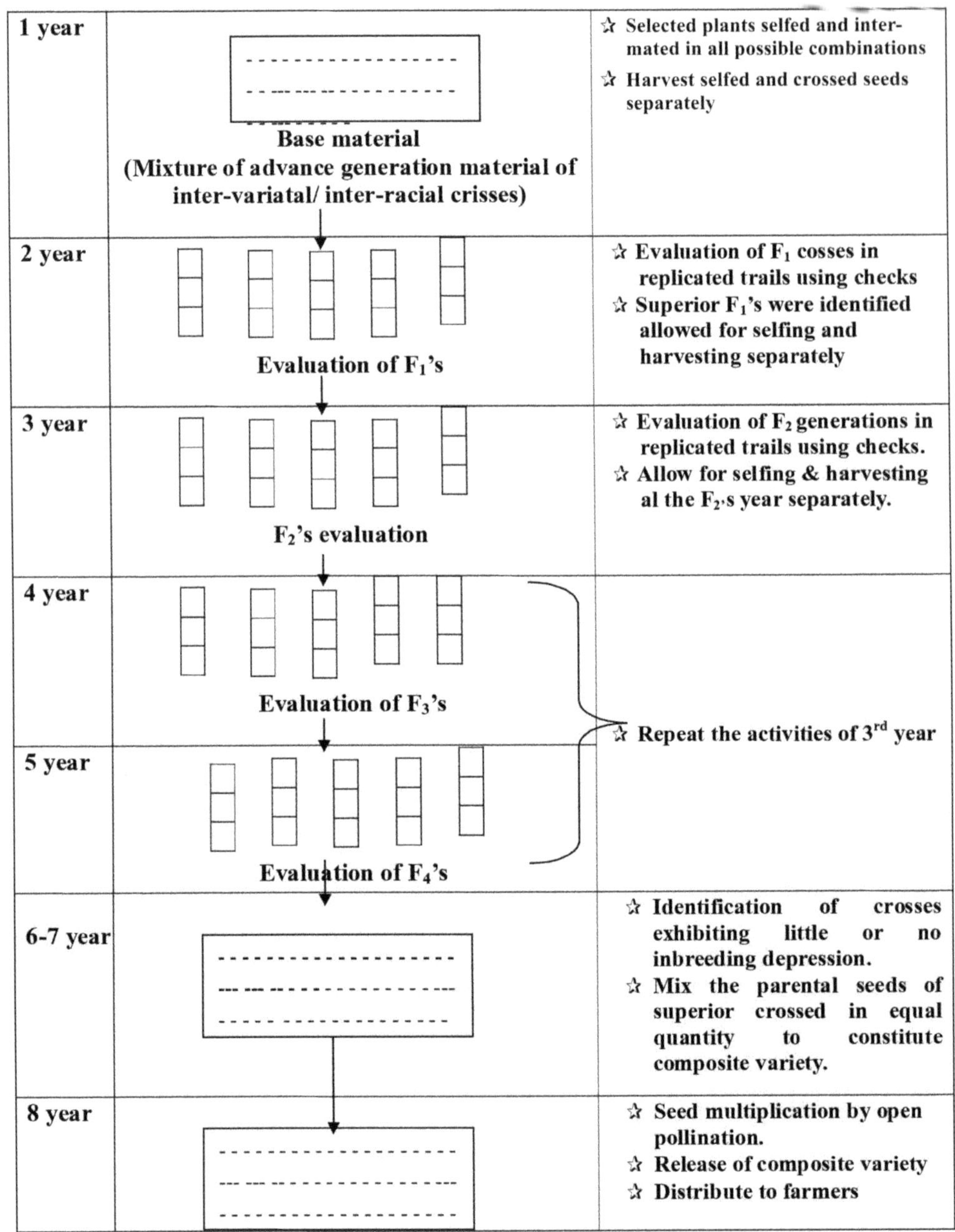

Figure 34: Steps Involved in Development of Composite Variety.

Table 34: Difference Between Synthetics and Composites

Synthetics	*Composites*
1. No. of inbredlines are less (6- 8)	No. of lines are more (even upto 20)
2. GCA of parental lines is tested	No. testing
3. Performance can be predicted	Cannot be predicted
4. Broad based	More broad based
5. Synthetic can be reconstituted	Cannot be reconstituted at a later date
6. Seed replacement after 4-5 years	After 3-4 years

15

Breeding for Vegetatively Propagated Crops

Some agricultural crops and a large number of horticultural crops are vegetative propagated.

Crops: Sugarcane, Potato, Sweet Potato, Tapioca, Ginger, Turmeric, Banana, *etc.*

Trees: Mango, Citrus, Apples, Pears, *etc.*

Characteristics of Asexually Propagated Crops

1 Majority of them are perennials: *e.g.* Sugarcane, fruit trees.

   The annual crops are mostly tuber crops: *e.g.* potato, cassava, sweet potato, *etc.*

2. Many of them show reduced flowering and seed set.
3. They are invariably cross pollinated.
4. These crops are highly heterozygous and show severe inbreeding depression upon selfing.
5. Majority of asexually propagated crops are polyploids: *e.g.* Sugarcane, potato, sweet, potato, *etc.*
6. Many species are inter-specific hybrids: *e.g.* Banana, sugarcane, *etc.*

Problems in Breeding Asexually Propagated Crops

1. Reduced flowering and fertility: This acts as a barrier in making crosses. Further apomixis and parthenocarpy prevents getting sexual progeny from crosses for further selection: *e.g.* mango

2. Lack of seed: *e.g.* Ginger, turmeric
3. There is short viability of seed: *e.g.* Sugarcane.
4. The seed production is very rare: *e.g.* Banana.
5. Seeds are produced under special conditions only: *e.g.* Sugarcane, potato, *etc.*
6. Difficulties in genetic analysis: Estimation of GCA and SCA are very much limited in clonally propagated crops. Without genetic analysis successful crop improvement is very much limited.
7. Most of their crops are perennials: Because of this regular yield trials with replication are not possible.

Breeding Methods

The vegetatively propagated crops can be improved by using following techniques:

1. Clonal Selection
2. Hybridization and Selection
   a) Inter-varietal
   b) Inter-specific
3. Polyploidy Breeding
4. Mutation Breeding
5. Tissue culture and Anther culture

Clone

A clone is a group of plants produced from a single plant through asexual reproduction. The crop plants can either be propagated by seeds or by vegetative parts.

Characteristics of Clones

1. All the individual belonging to a single clone are identical in genotype.
2. The phenotypic variation within a clone in due to environment only.
3. The phenotype of a clone is due to the effects of genotype (g), the environment (e) and the genotype x environment interaction (G x E).
4. Theoretically clones are immortal. They deteriorate due to viral/bacterial infection and mutations.
5. Clones are highly heterozygous and stable.
6. They can be propagated generation after generation without any change.

Importance of a Clone

1. Owing to heterozygosity and sterility in many crops clones are the only means of propagation.
2. Clones are used to produce new varieties.

3. Clones are very useful tools to preserve the heterozygosity once obtained. In many crops the superior plants are maintained. *e.g.* mango, orange, apple, sugarcane, *etc.*

Sources of Clonal Selection

1. Local varieties
2. Introduced material
3. Hybrids
4. Segregating populations

Clonal Degeneration

The loss in vigour and productivity of clones with time is known as clonal degeneration and results due to:

1. Mutation
2. Viral diseases
3. Bacterial diseases
4. Sexual reproduction
5. Mechanical mixture

Clonal Selection

The various steps involved in clonal selection are briefly mentioned below.

- ☆ **First Year:** From a mixed variable population, few hundreds to few thousands desirable plants are selected. Rigid selection can be done for simply inherited characters with high heritability. Plants with obvious weakness are eliminated.
- ☆ **Second Year:** Clones from the selected plants are grown separately, generally without replication. This is because of the limited supply of propagating material for each clone, and because of the large number of the clones involved. Characteristics of the clones will be clearer now than in the previous generation. Based on the observations the inferior clones are eliminated. The selection is based on visual observations and on judgment of the breeder on the value of clones. Fifty to one hundred clones are selected on the basis of clonal characteristics.
- ☆ **Third Year:** Replicated preliminary yield trial is conducted. A suitable check is included for comparison few superior performing clones with desirable characteristics are selected for multilocation trials. At this stage, selection for quality in done. If necessary, separate disease nurseries may be planted to evaluate disease resistance of the clones.
- ☆ **Fourth to Sixth Years:** Preliminary yield trials are conducted at several locations along with suitable check. The yielding ability, quality and disease resistance *etc.* of the clones are rigidly evaluated. The best clones

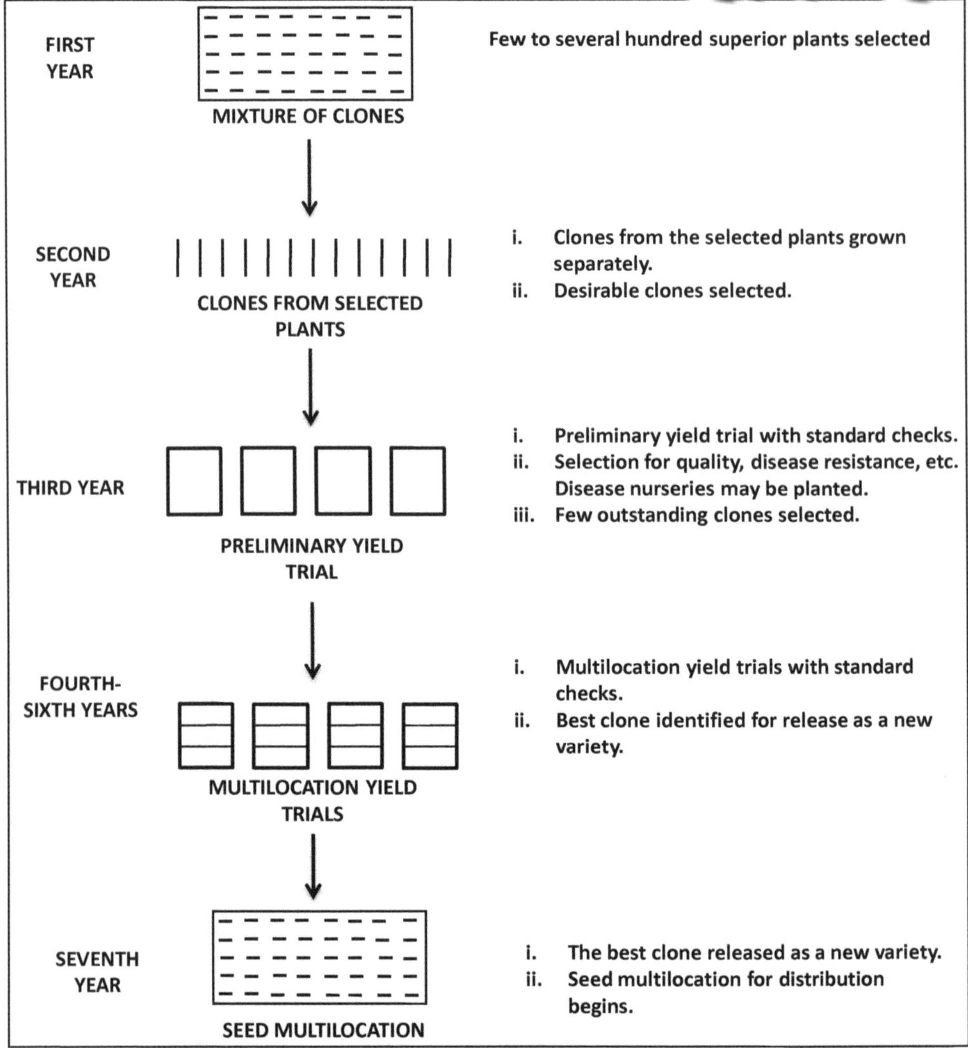

Figure 35: A Generalized Scheme for Clonal Selection in Asexually Propagated Species. This figure applies to a crop in which one generation does not take more than one year.

that are superior to the check in one or more characteristics are identified for release as varieties.

☆ **Seventh Year:** The superior clones are multiplied and released as varieties.

Genetic Variation within a Clone

1. **Mutation:** The frequency is generally very low (10-5 to 10-7). Ordinarily dominant mutations would be expressed in the somatic tissue. A mutant allele would be homozygous only when.

i) Both the alleles in a cell mutate at the same time producing the same mutant allele or

ii The mutant allele is already in heterozygous condition in the original clone. Though rare, both these events are possible. Bud mutations may often produce chimera's *i.e.* individual containing cells of two or more genotypes.

But mutations make possible selection of buds to establish new desirable clones, the process being known as Bud selection. It is of some importance in improvement of perennial crops like fruit trees or of those crops where flowering does not take place. It requires large number of plants to be observed and several trained persons to detect the mutant buds. Hence the bud selections are practiced in commercial plantations.

2. **Mechanical Mixtures:** Mechanical mixtures produce genetic variation within a clone much in the same way as in the case of purelines.
3. **Sexual Reproduction:** Occasional sexual reproduction would lead to segregation and recombination. The seedlings obtained from sexual reproduction would be genotypically different from the asexual progeny. It is evident that only clones would tend to became variable at least in annuals and biennials. *e.g.* Potato

Advantages

1. Varieties are stable and easy to maintain.
2. Avoids inbreeding depression.
3. Clonal selection, combined with hybridization generates necessary variability for several selections.
4. Only method to improve clonal crops.
5. Hybrid vigour is easily utilized selection may be used in maintaining the purity of clones.

Disadvantages

1. Selection utilizes the natural variability already present in the population.
2. Sexual reproduction is necessary for creation of variability through hybridization.
3. Applicable only to the vegetatively propagative crops.

Achievements

I. Through Clonal Selection

☆ **Potato:** 1. Kufri Red from Darjeeling Red Round

2. Kufri Safed from phulwa

☆ **Banana:** Bombay Green banana is a bud selection from dwarf Cavendish

II. Through Hybridization

- **Potato :** Kufri Alankar, Kufri Kuber, Kufri Sindhuri, Kufri Kundan, Kufri Chamatkar, Kufri Jyothi (late blight resistant), Kufri Sheetman (frost resistant), *etc.*
- **Sugarcane:** Co 1148, Co 1158, Co S 510, Co 975, Cos 109, Co 54, *etc.*
- **Mango:** Pedda Neelam, Chinna Suwarnarekha, *etc.*
- **Banana:** High gate from Gross Michel.
- **Citrus:** Robertson Navel Orange
- **Sweet oranges:** Yuvaraj blood Red
- **Turmeric**: Kesari, Kasturi, *etc.*

16

Heterosis and Inbreeding Depression

The term heterosis was first used by Shull in 1914. **Heterosis** may be defined as the superiority of an F_1 hybrid over both its parents in terms of yield and some other character. Generally, heterosis is manifested as an increase in vigour, size, growth rate, yield or some other characteristic. But in some cases, the hybrid may be inferior to the weaker parent. This is also regarded as heterosis; Often the superiority of F_1 is estimated over the average of the two parents, or the mid-parent. If the hybrid is superior to the mid-parent, it is regarded as heterosis (average heterosis or relative heterosis). However, in practical plant breeding, the superiority of F_1, over mid-parent is of no use since; it does not offer the hybrid any advantage over the better parent. Therefore, average heterosis is of little or no use to the plant breeder. More generally, heterosis is estimated over the superior parent; such an estimate is sometimes referred to as **heterobeltiosis.** The term heterobeltiosis is not commonly used since most breeders regard this to be the only case of heterosis and refer to it as such *i.e.,* heterosis. In 1944, Powers suggested that the term heterosis should be used only when the hybrid is either superior or inferior to both the parents. Other situations should be regarded as partial or complete dominance. However, the commercial usefulness of a hybrid would primarily depend on its performance in comparison to the best commercial variety of the concerned crop species. In many cases, the superior parent of the hybrid may be inferior to the best commercial variety. In such cases, it will be desirable to estimate heterosis in relation to the best commercial variety of the crop; such an estimate is known as economic, standard or useful heterosis. Economic heterosis is the only estimate of heterosis, which is of commercial or practical value.

Heterosis and Hybrid Vigour

Hybrid vigour has been used as a synonym of heterosis. It is generally agreed that hybrid vigour describes only the superiority of hybrids over their parents, while heterosis describes other situations as well. But a vast majority of the cases of heterosis are cases of superiority of hybrids over their parents. The few cases where F_1 hybrids are inferior to their parents may also be regarded as cases of hybrid vigour in the negative directions. For example, many F_1 hybrids in tomato are earlier than their parents. Earliness in many crops is agriculturally desirable. It may be argued that the earliness of F_1 hybrids exhibits a faster development in them so that their vegetative phase is replaced by the reproductive phase more quickly than in their parents. Therefore, the use of heterosis and hybrid vigour as synonym seems to be reasonably justified.

Luxuriance

Luxuriance is the increased vigour and size of inter-specific hybrids. The principal difference between heterosis and luxuriance lies in the reproductive ability of the hybrids. Heterosis is accompanied with an increased fertility, while luxuriance is expressed by inter-specific hybrids that are generally sterile or poorly fertile. In addition, luxuriance may not result from either masking of deleterious genes or from balanced gene combinations brought together into the hybrid. Therefore, luxuriance does not have any adaptive significance.

Selected Hisorical Milesotens

Hybrid vigour in artificial tobacco (*Nicotiana species*) hybrids was first reported by **Koelreuter** in 1673. Subsequently, many workers reported hybrid vigour in a large number of plant species. These hybrids were produced from inter-specific as well as intra-specific crosses. In 1876, **Darwin** concluded that hybrids from unrelated plant types were highly vigorous. Most of our present knowledge on heterosis comes from the work on maize. Maize is perhaps the most extensively studied crop species with respect to heterosis and inbreeding depression. **Beal** studied the performance of inter-varietal hybrids between 1877 and 1882. He reported that some hybrids yielded as much as 40 per cent more than the parental varieties. From subsequent studies on inter-varietal crosses in maize, it became clear that some of the hybrids showed heterosis, while others did not. Crosses between distinct types, *i.e.,* genetically diverse varieties, exhibited greater heterosis than those involving closely related varieties.

Heterosis in Cross - and Self-pollinated Species

In general, cross-pollinated species show heterosis, particularly when inbred lines are used as parents. In many cross-pollinatedspecies, heterosis has been commercially exploited, for example, in maize, bajra, jowar, cotton, sunflower, onion, alfalfa, *etc.* Many crosses in self-pollinated species also show heterosis, but the magnitude of heterosis is generally smaller than that in the case of cross-pollinatedspecies. But in some self-pollinated crops, heterosis is large enough to be used for the production of hybrid varieties. Hybrid varieties are commercially used in some vegetables, such as tomato, where a single fruit produces a large number of seeds, and in crops like rice. The chief drawback in the use of hybrid varieties in self-pollinated crops is the great difficulty encountered in the production of large quantities of hybrid seed.

Manifestations of Heterosis

These various manifestations of heterosis may be summarized as follows.

1. **Increased yield:** Heterosis is generally expressed as an increase in the yield of hybrids.
2. **Increased Reproductive Ability:** The hybrids exhibiting heterosis show an increase in fertility or reproductive ability. This is often expressed as higher yield of seeds or fruits or other propagules, *e.g.*, tuber in potato (*S. tuberosum*), stem in sugarcane (S. *officinarum*), *etc.*
3. **Increase in Size and General Vigour:** The hybrids are generally more vigorous, *i.e.*, healthier and faster growing and larger in size than their parents. *Some examples* of increased size are increases in fruit size in tomato, head size in cabbage, cob size in maize, head size in jowar, *etc.*
4. **Better Quality:** *For example*, many hybrids in onion show better keeping quality, but not yield, than open- pollinated varieties.
5. **Earlier Flowering and Maturity:** In many cases, hybrids are earlier in flowering and maturity than the parents. This may sometimes be associated with a lower total plant weight. But earliness is highly desirable in many situations, particularly in vegetables. Many tomato hybrids are earlier than their parents.
6. **Greater Resistance to Diseases and Pests:** Some hybrids are known to exhibit a greater resistance to insects or diseases than their parents.
7. **Greater Adaptability:** Hybrids are generally more adapted to environmental changes than inbreds. In general, the variance of hybrids is significantly smaller than that of inbreds. This shows that hybrids are more adapted to environmental variations than are inbreds. In fact, it is one of the physiological explanations offered for heterosis.
8. **Faster Growth Rate:** In some cases, hybrids show a faster growth rate than their parents. But the total plant size of the hybrids may be comparable to that of parents. In such cases, a faster growth rate is not associated with a larger size.
9. **Increase in the Number of a Plant Part:** In some cases, there is an increase in the number of nodes, leaves and other plant parts, but the total plant size may not be larger.

These are some of the characteristics for which heterosis is easily observed. Many other characters are also affected by heterosis, *e.g.* enzyme activities, cell division, vitamin content (vitamin C content in tomato), other biochemical characteristics, *etc.*, but they are not so readily observable.

Genetic Basis of Heterosis and Inbreeding Depression

Heterosis and inbreeding depression are closely related phenomena. In fact, they may be regarded as the opposite sides of the same coin. Therefore, genetic theories that explain heterosis also explain inbreeding depression. There are three main theories to explain heterosis and, consequently, inbreeding depression: (1) Dominance, (2) Over dominance, and (3) Epistatis hypotheses.

Dominance Hypothesis

The dominance hypothesis was first proposed by **Davenport** in 1908. It was later expanded by Bruce and by Keeble and Pellew in 1910. In simplest terms, this hypothesis suggests that at each locus the dominant allele has a favourable effect, while the recessive allele has an unfavourable effect. In heterozygous state, the deleterious effects of recessive alleles are masked by their dominant alleles.

Thus heterosis results from the masking of harmful effects of recessive alleles by their dominant alleles. Inbreeding depression, on the other hand, is produced by the harmful effects of recessive alleles, which become homozygous due to inbreeding. Therefore, according to the dominance hypotheses, heterosis is not the result of heterozygosity; it is the result of prevention of expression of harmful recessives by their dominant alleles. Similarly, inbreeding depression does not result from homozygosity per se. but from the homozygosity of recessive alleles, which have harmful effects.

Objections

1. **Failure in the Isolation of Inbreds as Vigorous as Hybrids.** According to the dominance hypothesis, it should be possible to isolate inbreds with all the dominant genes. Such inbreds would be as vigorous as the F_1 hybrids. However, such inbreds have not been isolated in many studies. But in some studies, it has been possible to recombine genes so that inbred lines as good as or superior to the heterotic hybrids were isolated.
2. **Symmetrical Distribution in F_2:** In F_2, dominant and recessive characters segregate in the ratio of 3:1. According to the dominance hypothesis, quantitative characters, therefore, should not show a symmetrical distribution in F_2. This is because dominant and recessive phenotypes would segregate in the proportion $(\frac{3}{4} + \frac{1}{4})^n$, where n is the number of genes segregating. However, F_2's nearly always show a symmetrical distribution.

Over-dominance Hypothesis

This hypothesis was independently proposed by **East and Shull** in 1908. This is sometimes known as single gene heterosis, super dominance, cumulative action of divergent alleles, and stimulation of divergent alleles. The idea of super dominance, *i.e.*, heterozygote superiority, was initially put forth by Fisher in 1903; it was elaborated by East and Shull in 1908 to explain heterosis. According to over dominance hypothesis, heterozygote's at least some of the loci are superior to both the relevant homozygote's. Thus heterozygote *Aa* would be superior to both the homozygote's *AA* and *aa*. Consequently, heterozygosity is essential for and is the cause of heterosis, while homozygosity resulting from inbreeding produces inbreeding depression. It would, therefore, be impossible to isolate inbreds as vigorous as F_1 hybrids if heterosis were the consequence of over dominance.

Evidence for Over-dominance

There are not many clear- cut cases where the heterozygote is superior to the two homozygote's; in fact, over dominance has not been demonstrated unequivocally for any polygenic trait (Banga and Banga, 1998). This has been the biggest objection to the general acceptance of over dominance hypothesis. But there is no doubt that in the case of some oligogenes, heterozygotes are superior to the homozygotes. In case of maize, gene ma affects maturity. The heterozygote *Ma ma* is more vigorous and later in anthesis and maturity than the homozygotes *Ma Ma* and *ma ma*.

Table 35: A Comparison between Dominance and Overdominance Hypotheses of Heterosis

Feature	*Hypothesis of Heterosis*	
	Dominance	*Overdominance*
Similarities		
Inbreeding leads to leads to	Reduced vigour and fertility	Reduced vigoour and fertility
Out-crossing leads to	Heterosis	Heterosis
Degree of heterosis increases with	Genetic diversity between parents	Genetic diversity between parents
Differences		
Inbreeding depression is the results of	Homozygosity for deleterious recessive alleles	Homozgosity itself
Heterosis is the result of	Making of the harmful effects of recessive alleles by their dominant alleles	Heterozogosity itself
The phenotype of heterozygote is	Comparable to that of the dominant homozygote	Superior to both the homozygotes
Inbreds as vigorous as the F_1 hybrid	Can be isolated	Cannot be isolated

Epistasis Hypothesis

In 1952, ***Gowen*** had suggested that influence of one locus on the expression of another may be involved in heterosis. Subsequently, considerable data has accumulated to implicate epistasis as a cause of heterosis. *For example,* a majority of heterotic crosses show significant epistasis. But all heterotic crosses do not show epistasis, and all crosses that show epistasis are not heterotic. In many cases, the effect of a single homozygous recessive allele is epistatic to almost the whole genetic makeup of an inbred. When the effects of such an allele are masked by its dominant allele, the effects on heterosis are usually dramatic (Stuber 1994). However, epistatic variance usually forms only a much smaller component of the total genetic variance than do additive and dominance variances.

Spite of the large experimental evidence accumulated, it is not possible to conclusively accept or reject one or the other hypothesis. There are definitely some genes that show heterozygote superiority. But the number of such genes appears to be rather small, and even these cases could be due to linkage in repulsion phase or epistasis or both. It is generally accepted that heterosis, to a large extent, is due to dominance gene action, but epistasis and over dominance are also involved (both in self- and cross -pollinated crops).

Estimation of Heterosis

Heterosis is estimated in three different ways, *viz.*,

1. Mid-Parent Heterosis Average Heterosis/Relative Heterosis

It refers to the superiority of an F1 over the mid parent value *i.e.* average of the two parental means. It is also called as relative heterosis and is estimated as follows.

$$\text{Mid - parent heterosis} = [(\overline{F_1}\quad \overline{MP})/\overline{MP}] \times 100$$

Where, $\overline{F_1}$ is $\overline{MP}$ the mean value of F_1 and is the mean value of two parents involved in the cross.

2. Heterobeltiosis

It refers to the superiority of an F_1 over its better parent. It is estimated as follows.

$$\text{Heterosis} = [(\overline{F_1}\quad \overline{BP})/\overline{BP}] \times 100$$

Where, $\overline{BP}$ is the mean value (over replications) of the better parent of a cross

3. Useful/Standard Heterosis/Economic Heterosis

The term useful heterosis was used by Meredith and Bridge (1972). It refers to the superiority of F_1 over the standard commercial check variety. It is also called as economic heterosis. It is estimated as follows.

$$\text{Usefulheterosis} = [(\overline{F_1}\quad \overline{CC})/\overline{CC}] \times 100$$

Where, $\overline{CC}$ is the mean value over replications of the local commercial cultivar.

Sometimes heterosis is worked out over the standard commercial hybrid. It is estimated in those crops where hybrids are already available for comparison. It is estimated as follows.

$$\text{Heterosis} = [(\overline{F_1}\quad \overline{SH})/\overline{SH}] \times 100$$

Where, $\overline{SH}$ is the mean value over replications of the local commercial hybrid.

Inbreeding is mating of individual more closely related by ancestry than would be expected under random mating. In other words, it is mating between individuals related by ancestry or descent and the highest degree of inbreeding is achieved by selfing. The chief effect of inbreeding is an increase in homozygosity in the progeny. In fact the degree of inbreeding in any generation is equal to the degree of homozygosity in that generation.

Loss in vigour due to inbreeding is known as inbreeding depression or IBD may be defined as the reduction or loss in vigour and reproductive capacity as a result of inbreeding. Inbreeding generally results in accumulation and appearance of lethal and sub lethal recessive alleles in progeny, causing IBD. The IBD can be calculated by using following formula:

$$IBD = \frac{\overline{F_2}\ \overline{F_1}}{\overline{F_1}} \times 100$$

Inbreeding Coefficient is given by F= 1-(1/2)^{n}, Where n: No. of offsprings.

Historical

Inbreeding depression has been recognized by man for a long time. A marriage between closely related individuals has been prohibited since early time in many societies. Because people aware of the harmful effects of such marriages in the progeny. A systematic observation on effect of inbreeding started during 17th century when inbreeding became a common practice in cattle breeding. In 1876, **Darwin** published his book on **cross and self-fertilization in vegetable kingdom**. He concluded that progeny obtained from self fertilization were weaker than those obtained from out crossing. **East (1908) and Shull (1909)** independently showed the effect of inbreeding depression while working in maize. Subsequently scientists reported inbreeding depression in other crop plants.

Effects of Inbreeding

The various plant species differ considerably in their response to inbreeding and any range from very high to very low or absent. It has become clear that in cross-pollinatedand in asexually propagated species, inbreeding has harmfull effects which are often severe. The different effects of inbreeding are:

1. **Appearance of Lethal and Sub lethal Alleles**: Inbreeding results in appearance of lethal; sub lethal and sub vital characters. *E.g*: Chlorophyll deficiencies, rootless seedlings, flower deformities (they do not survive, they lost in population).
2. **Reduction in Vigour**: General reduction in vigour size of various plant parts.
3. **Reduction in Reproductive ability**: Reproductive ability of population decreases rapidly. Many lines reproduce purely that they cannot be maintained.
4. **Separation of the population into distinct lines**: Population rapidly separates into distinct lines *i.e.* due to increase in homozygosity. This leads to random fixation of alleles in different lines. Therefore lines differ in genotype and phenotype. It leads to increase in the variance of the population.
5. **Increase in Homozygosity**: Each line becomes homozygous. Therefore, variation within a line decreases rapidly. After 7-8 generations of selfing the line becomes more than 99 per cent homozygous. These are the inbreds.

6. **Reduction in Yield**: Inbreeding leads to loss in yield. The inbreds that survive and maintained have much less yield than the open pollinated variety from which they have been developed.

Degrees of Inbreeding Depression

1. **High Inbreeding Depression**: *e.g*, Alfalfa and carrot show very high ID. A large proportion of plants produced by selfing show lethal characteristics and do not survive.
2. **Moderate Inbreeding Depression**: *e.g*, Maize, Jowar and Bajra *etc.* show moderate ID.
3. **Low Inbreeding Depression**: *e.g.*, Onion, many Cucurbits, Rye and Sunflower *etc.* show a small degree of ID.
4. **Lack of inbreeding depression**: The self-pollinated species do not show ID, although they do show heterosis. It is because these species reproduce by self-fertilization and as a result, have developed homozygous balance.

Inbreeding Depression and Heterosis Breeding

These two aspects are of great significant in breeding cross pollination crops. Most of the improved varieties in cross pollination crops are either hybrids or composites.

It is clear that heterotic advantage is limited only to the F_1 generation and thereafter it declines. Thus, for the maximum utilization of heterosis fresh hybrid seeds have to be produced and used every year.

Table 36: A List of Plant and Animal Species where Heterosis is Being Commercially Exploited

Category	*Examples*
Crop species	1. Asexually propagated species 2. Cross-pollinated species: maize, jowar, bajra, sugarbeets, sunflower, forage grasses, castor, forage legumes and cotton 3. Self-pollinated crops: rice, pigeonpea (India)
Vegetable crops	Tomato, brinjal (Solanum melongena), onion, Brussel's sprouts, Watermelon, pepper, winter sqash, muskmelon, cabbage, broccoli, spinach, red beets, carrot cauliflower, celery, asparagus
Fruit trees	In almost all the fruit trees
Animals	Silkworm, poultry, cattle, swine

Mutagenesis in Plant Breeding

It was previously pointed out that mutation is the ultimate source of variation. Without adequate variation, plant breeding is impossible. To start a breeding program, the breeder must find the appropriate genotype (containing the desired genes) from existing variation, or create the variation if it is not found in nature. Mutagenesis is the process by which new alleles are created. **Mutation** is the sudden heritable change in specific characteristic an organism.

Brief Historical Perspective

- The term mutation was coined by **Hugo de Veries** (1900), for the first time and the word is derived from the Latin word *'MUTARE'* means to change. He observed sudden heritable change in evening prime rose (*Oenothera lemarkiana*). He called these changes as **mutation** and organism undergone mutation were termed as **mutants** (bearing mutant gene).
- The scientific study of mutation was started in 1910 by **Morgan and his workers** in *Drosphila*. He observed white eyed male among red eyed male individuals. **The white eyed male was a mutant.**
- The discovery of the mutagenic effects of X-rays on the fruit fly (*Drosophila*) by **H. Muller** in the 1927 paved the way for researchers to experiment with its effects on various organisms and developed CIB to detect the mutations.
- In 1928, **H. Stubbe** demonstrated the use of **mutagenesis** in producing mutants in tomato, soybean, and other crops.

- In 1928 Stadler used gamma rays and x- rays as mutagens in barley and maize.
- 1930: Charles Aureback (grand lady of chemical mutagenesis) first to find out chemical mutagens.
- The first commercial mutant **Chlorina** was produced in tobacco in 1934 by Tollenuar (Dutch).
- Ake Gustafsson (1908-88): Identified chlorophyll mutants (1940) and developed Pallus variety in barley by treating with X- rays (1960).
- 1946: Auerbach and Robson reported that nitrogen mustard produced mutation in Drosophila.
- 1950: B. Mc Clintock demonstrated mutation by X- rays and also identified jumping gene (transposable elements) in Maize.
- Currently, induced mutations are used more often in a supplementary role as a source of new alleles.
- However, it is still important in breeding vegetatively propagated species, including field crops, ornamentals, and fruit and forest species.
- It is especially useful in ornamental plant breeding where novelty is often advantageous and can become commercially significant.
- Furthermore, with the advent of genetic engineering and its radical tools, which allow targeted genetic alteration (versus the random genetic alteration produced by conventional mutagenesis), it appears that breeders are gravitating towards this truly revolutionary technology for creating new variability.

In conventional breeding of sexual plants, genetic variability is derived from recombination. Parents must not be identical, or else there would be no segregation in the F_2 generation. Even when parents are dissimilar, they often have similar *"house keeping genes"* that are common to both parents. Whereas segregation will not occur for these common genes, mutagenesis can create variability by altering them.

Characteristic Feature of Mutations

1. Mutations are generally recessive but dominant mutations also occur.
2. Mutations are generally harmful to the organism. Most of the mutations have deleterious effects but small proportions (0.1 per cent) of them are beneficial.
3. Mutations are random *i.e.* they may occur in any gene. However some genes show high mutation rates than the others.
4. Mutations are recurrent.
5. Induced mutations commonly show pleiotropy often due to mutation in closely linked genes.

Classification of Mutations

1. Somatic vs. Gametic Mutations

The consequences of a mutation depend upon where in an individual they occur. Some mutations occur in regular body cells; these are **somatic** mutations. For example, someone who spends too much time sun tanning might experience a mutation in a skin cell. The consequences of such a mutation are felt only by the individual. The skin cell may develop some problem (such as cancer, perhaps) as a result of the mutation, but because the mutation occurred only in a skin cell, it would not be passed on to subsequent generations.

Some mutations occur in germline cells. These cells produce the gametes; therefore, they are **gametic** mutations. In most cases, such mutations wouldn't even be noticed by the individual. After all, the gametes don't play a prominent role in the day-to-day function of the individual. These mutations, in contrast to the somatic mutations, will be passed on to the next generation, because they occur in the cells that produce the next generation.

2. Spontaneous vs. Induced Mutations

Some mutations arise as natural errors in DNA replication (or as a result of unknown chemical reactions); these are known as **spontaneous** mutations. The rates of such mutations have been determined for many species. *E. coli* has a spontaneous mutation rate of $1/10^8$ (one error in every 10^8 nucleotides replicated). Humans have a higher spontaneous mutation rate: between $1/10^6$ and $1/10^5$ (probably as a result of the higher complexity of human replication).

Mutations can also be caused by agents in the environment; these are **induced** mutations. Induced mutations increase the mutation rate over the spontaneous rate. Looking at a single mutation in an individual, one cannot tell if the mutation was spontaneous or induced. Induced mutations can only be described by looking at the mutation rate in a population, and comparing it to the spontaneous mutation rate for the species. If the observed mutation rate is higher, then induced mutations can be assumed. Agents in the environment that cause an increase in the mutation rate are called **mutagens**.

3. By Effect on Structure

The sequence of a gene can be altered in a number of ways. Gene mutations have varying effects on health depending on where they occur and whether they alter the function of essential proteins. Mutations in the structure of genes can be classified as:

Small-scale Mutations

Such as those affecting a small gene in one or a few nucleotides, including:

- ☆ Point mutations, often caused by chemicals or malfunction of DNA replication, exchange a single nucleotide for another. These changes are classified as transitions or transversions. Most common is the transition that exchanges a purine for a purine (A/G) or a pyrimidine for a pyrimidine, (C/T). Less common is a transversion, which exchanges a purine for a pyrimidine or a pyrimidine for a purine (C/T ! A/G). An example of a transversion is adenine (A) being converted into a cytosine (C).

- Insertions add one or more extra nucleotides into the DNA. They are usually caused by transposable elements, or errors during replication of repeating elements (*e.g.* AT repeats).
- **Deletions** remove one or more nucleotides from the DNA. Like insertions, these mutations can alter the reading frame of the gene.

Large-scale Mutations in Chromosomal Structure

- **Amplifications** (or gene duplications) leading to multiple copies of all chromosomal regions, increasing the dosage of the genes located within them.
- **Deletions** of large chromosomal regions, leading to loss of the genes within those regions.
  - Chromosomal translocations: interchange of genetic parts from non homologous chromosomes.
  - **Interstitial deletions**: an intra-chromosomal deletion that removes a segment of DNA from a single chromosome, thereby apposing previously distant genes.
  - Chromosomal inversions: reversing the orientation of a chromosomal segment.
- Loss of heterozygosity: loss of one allele, either by a deletion or recombination event, in an organism that previously had two different alleles.

4. By Effect on Function

- **Loss-of-function mutations** are the result of gene product having less or no function. When the allele has a complete loss of function (null allele) it is often called an amorphic **mutation**.
- **Gain-of-function mutations** change the gene product such that it gains a new and abnormal function. These mutations usually have dominant phenotypes. Often called a neomorphic mutation.
- **Dominant negative mutations** (also called antimorphic **mutations**) have an altered gene product that acts antagonistically to the wild-type allele. These mutations usually result in an altered molecular function (often inactive) and are characterized by a dominant or semi-dominant phenotype.
- **Lethal mutations** are mutations that lead to the death of the organisms which carry the mutations.

☆ A **back mutation** or **reversion** is a point mutation that restores the original sequence and hence the original phenotype.

5. By Effect on Fitness

☆ **Harmful mutation** is a mutation that decreases the fitness of the organism.

☆ **Beneficial mutation** is a mutation that increases fitness of the organism, or which promotes traits that are desirable.

☆ **Neutral mutation** has no harmful or beneficial effect on the organism. Such mutations occur at a steady rate, forming the basis for the molecular clock.

☆ **Deleterious mutation** has a negative effect on the phenotype, and thus decreases the fitness of the organism.

☆ **Advantageous mutation** has a positive effect on the phenotype, and thus increases the fitness of the organism.

☆ **Nearly neutral mutation** is a mutation that may be slightly deleterious or advantageous, although most nearly neutral mutations are slightly deleterious.

6. By Impact on Protein Sequence

☆ A frameshift mutation is a mutation caused by insertion or deletion of a number of nucleotides that is not evenly divisible by three from a DNA sequence. Due to the triplet nature of gene expression by codons, the insertion or deletion can disrupt the reading frame, or the grouping of the codons, resulting in a completely different translation from the original. The earlier in the sequence the deletion or insertion occurs, the more altered the protein produced is.

☆ A nonsense mutation is a point mutation in a sequence of DNA that results in a premature stop codon, or a *nonsense codon* in the transcribed mRNA, and possibly a truncated, and often nonfunctional protein product.

☆ **Missense mutations** or *non synonymous mutations* are types of point mutations where a single nucleotide is changed to cause substitution of a different amino acid. This in turn can render the resulting protein nonfunctional. Such mutations are responsible for diseases such as *Epidermolysis bullosa*, sickle-cell disease, *etc.*

☆ A neutral mutation is a mutation that occurs in an amino acid codon which results in the use of a different, but chemically similar, amino acid. The similarity between the two is enough that little or no change is often rendered in the protein. For example, a change from AAA to AGA will encode lysine, a chemically similar molecule to the intended arginine.

☆ **Silent mutations** are mutations that do not result in a change to the amino acid sequence of a protein. They may occur in a region that does not code for a protein, or they may occur within a codon in a manner that does not alter the final amino acid sequence. The phrase *silent mutation* is often used interchangeably with the phrase *synonymous mutation*; however, synonymous mutations are a subcategory of the former, occurring only within exons. The name silent could be a misnomer. *For example*, a silent mutation in the exon/intron border may lead to alternative splicing by changing the splice site (*see Splice site mutation*), thereby leading to a changed protein.

Tautomerism

- When a molecule is able to exist in more than one chemical form, it is called **tautomeric** and the phenomenon is known as **tautomerism.** Radiation may provide the energy for the formation of tautomeric forms.
- Normally **Adenine**: Links with **Thyamine and Guanine**: Links with **Cytosine**.
  - These are the base paring found in DNA molecule, but due to tautomerism an unusual base pairing, such as **A – C or G – T** may result. Such unusual base paring always, cause changes in the character of the progeny.

Bud Mutation

If the mutation occurs in the meristematic in the early stages of bud development, all the cells of the bud will be mutant in nature, to the shoot developed from such bud will be a mutant one. This type of mutation will be called as bud sport. If a mutation occurs in the last stages of bud development, only some of the cells of the bud will be mutant in nature. A plant which has genetically distinct tissues lying adjacent to one another is called a ***chimera.***

- **Chimeras classified into 3 types:**
  1. **Periclinal chimera**: Mutant and non-mutant tissues are in concentric layers one overlapping the other. They can be perpetuated by vegetative propagation.
  2. **Sectorial chimera**: A segment of mutant tissue extending from the epidermis towards the centre.
  3. **Mericlinal:** It is incomplete periclinal chimera. The mutant tissues partly surround Mutant.

Mutagenic Agents

Agents of artificial mutations are called **mutagens**. They may be grouped into two broad categories *viz.*, **physical mutagens** and **chemical mutagens**. The specific agents vary in ease of use, safety issues, and effectiveness in inducing certain genetic alterations, suitable tissue, and cost, among other factors.

Table 37: List of Physical Mutagens and their Important Features

Mutagen	*Characteristics*
X-rays	Electromagnetic radiation; penetrate tissues from a few millimeters to many centimeters
Gamma rays	Electromagnetic radiation produced by radioisotopes and nuclear reactors; very penetrating into tissues; sources are Co ^{60} and Ce^{137}
Neutrons	A variety exists (fast, slow, thermal); produced in nuclear reactors; uncharged particles; penetrate tissues to many centimeter; source is U^{235}
Beta particles	Derived from radioisotopes; a helium nucleus capable of heavy ionization; very shallowly penetrating
Protons	Produced in nuclear reactors and accelerators; derived from hydrogen nucleus; penetrate tissues up to several centimeters

Table 38: List of Chemical Mutagens and their Important Features

Mutagen Group	*Examples*
Base analogues	5 – bromouracil, 5-bromodeoxyuridine
Related compounds	Maleic hydrazide, 8-ethoxy caffeine
Antibiotics	Actinomcin D, mitomycin C, streptonigrin
Alkylating agents	
Sulfur mustards	Ethyl-2-chloroethyl sulfide
Nitrogen mustards	2- chloroethyl-dimethyl amine
Epoxides	Ethylene oxide
Ethyleneimines	Ethyleneimine
Sulfonates, *etc.*	Ethyl methane sulfonate (EMS), diethylsulfonate (DES)
Diazoalanes	Diazomethane
Nitroso compounds	N-ethyl-N-nitroso urea
Azide	Sodium azide
Hydroxylamine	Hydroxylamine
Nitrous acid	Nitrous acid
Acridines	Acridine orange

Factors Affecting the Success of Mutagenesis

Mutations are random events even when scientists induce them. The plant breeder using the technique can increase the success rate by observing the following cautions.

1. **Clear objective**: A program established to select one specific trait is more focused and easier to conduct with a higher chance of success, than a program designed to select more than one trait.
2. **Efficient screening methods**: Mutation breeding programs examine large segregating populations to increase the chance of finding the typically rare

desirable mutational events. An efficient method of screening should be developed for a mutation breeding program.

3. **Proper choice of mutagen and method of treatment**: Mutagens, as previously discussed, vary in various properties including source, ease of use, penetration of tissue, and safety. Some are suitable for soft tissues, whereas others are suited to hard tissue.
4. **Dose rate**: The breeder should decide on the appropriate and effective dose rate (dose and duration of application). The proper dose rate is determined by experimentation for each species and genotype. Plant materials differ in sensitivity to mutagenic treatment. It is difficult to find the precise dose (intensity), but careful experimentation can identify an optimum dose rate.
5. **Proper experimental conditions**: It is known that the oxygen level in the plant material affects the amount of damage caused by the mutagen – the higher the oxygen level, the greater the injury to the material tends to be. The change in the effect of a mutagen with oxygen supply is called the **oxygen enhancement ratio**.

Another experimental factor affecting the success of mutagenesis is the pH of the environment in chemical mutagenesis. *For example*, EMS is most effective at pH 7.0, whereas sodium azide is most effective at p^H 3.0. Sometimes, dry seeds need to be pre-soaked to prepare the cells to initiate metabolic activity. It is important to mention that it is easier to modify the experimental condition when seeds are being used for the mutation program than when other materials are used.

Source for Irradiation

The plant material may be treated in any of the following source.

- ☆ **Seeds:** Seeds are used after soaking to get greater frequency of induced mutations than air dried.
- ☆ **Seedlings :** At any stage of life cycle can be subjected to radiation but usually seedlings neither too young nor too old are irradiated due to their convenience in handling in pots transportation from nursery easily.
- ☆ **Flowers:** Meiotic cells have been found more sensitive than the mitotic cells and therefore plants are irradiated in the flowering stage in order to affect the developing gametes.
- ☆ **Cuttings:** In case of fruit tree when they are propagated by clones–the desirable cuttings are exposed to irradiation.

Selection of the Variety

Usually the locally adapted best variety will be selected. For *e.g.* for induction of male sterility in Red gram, locally adapted short duration TS 3R is to be selected. But this may not be Universal rule. For *e.g.* to breed alternate dwarfing gene in rice, low yielding tall *indicas* may be subjected to mutagenic treatments.

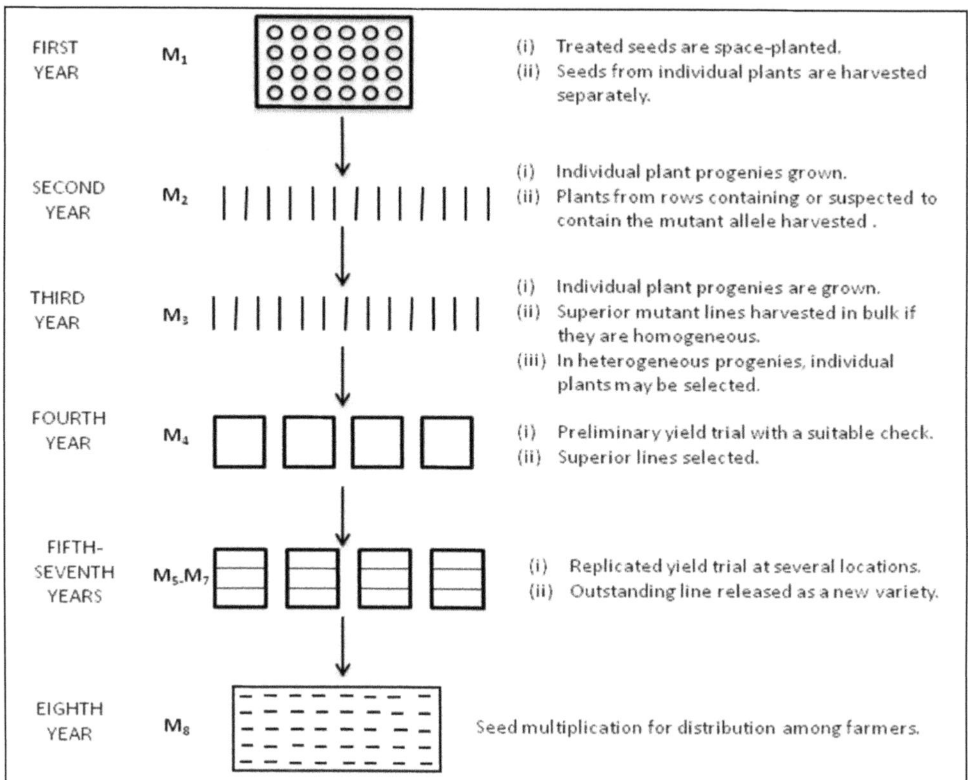

Figure 36: A Generalized Scheme for Mutation Breeding for an Oligogenic Trait; the mutant allele is recessive.

Procedure

The handling procedure described here is based on the selection for a recessive mutant allele of an oligogene.

1. **M_1:** Several hundred seeds are treated with a mutagen and are space planted. In general, the number of treated seeds is so adjusted as to give rise to 500 fertile M_1 plants at the harvest. Care should be taken to avoid out crossing; This can be achieved either by planting the M_1 population in isolation or by bagging the inflorescences of M_1 plants or even the whole M_1 plants. M_1 plants will be chimeras for the mutations present in heterozygous state. About 20 to 25 seeds from each M_1 spike are harvested separately to raise the M_2 progeny rows.
2. **M_2:** About 2,000 progeny rows are grown. Careful and regular observations are made on the M_2 rows. But only distinct mutations are detected in M_2 because the observations are based on single plants. All the plants in M_2 rows suspected of containing new mutations are harvested separately to raise individual plant progenies in M_3. If the mutant is distinct, it is selected for multiplication and testing. However, most of the mutations

will be useless for crop improvement. Only 1-3 per cent of M_2 rows may be expected to have beneficial mutations.

Alternatively, M_2 may be grown as a bulk produced by compositing one or more, but equal number of, seeds from each M_1 spike/fruit/branch. Individual plants are then selected in M_2 and individual plant progenies are grown in M_3.

3. **M_3:** Progeny rows from individual selected plants are grown in M_3. Poor and inferior mutant rows are eliminated. If the mutant progenies are homogeneous, two or more M_3 progenies containing the same mutation may be bulked. Mutant M_3 rows are harvested in bulk for a preliminary yield trial in M_4.
4. **M_4:** A preliminary yield trial is conducted with a suitable check, and promising mutant lines are selected for replicated multilocation trials.
5. **M_5-M_7:** Replicated multilocation yield trials are conducted. The outstanding line may be released as a new variety. The low yielding mutant lines, however, should be retained for use in hybridization programmes.

Applications of Mutation Breeding

1. Induction of desirable mutant alleles which may not be available in the germplasm.
2. It is useful in improving specific characteristics of a well adapted high yielding variety.
3. Mutagenesis has been successfully used to improve various quantitative characters including yield.
4. F_1 hybrids from inter-varietal crosses may be treated with mutagens in order to increase genetic variability by inducing mutation and to facilitate recombination of linked genes.
5. Irradiation of inter-specific (distant) hybrids has been done to produce translocations.
6. For induction of male sterility induced mutagenesis.
7. Mutation creates inexhaustible variation.

Limitations

1. Frequency of desirable mutations is very low about 0.1 percent.
2. To detect the desirable one in M_2 considerable time, labour and other resources are to be employed.
3. To screen large population, efficient quick and inexpensive selection techniques are needed.
4. Desirable mutations may be associated with undesirable side effects due to other mutations thus extending the mutation breeding programme.
5. Detection of recessive mutations in polyploids and clones is difficult.

6. Larger doses of mutagen have to be applied and larger populations are to be grown.
7. Hit or miss method.

Achievements

a) Natural Mutants

☆ **Rice**: GFB 24 – arose as a mutant from Konamani variety, Dee – Gee – Woo – Gen – Arose as a mutant from rice in China and MTU 20 – arose as a mutant from MTU-3

☆ **Sorghum**: Co. 18 – arose as a mutant from Co. 2

☆ **Cotton**: DB 3- 12 from *G. heroaccum* variety Western 1.

Table 39: Selected General Areas of Achievement in Mutation Breeding

Disease resistance: *e.g.*, *verticilium wilt* resistance in peppermint, victorial blight resistance in barley, downy mildew resistance in pearl millet
Modification of plant structure: *e.g.*, bush habit in dry bean, dwarf mutants in wheat and other cereals
Nutritional quality augmentation: *e.g.*, opaque and floury endosperm mutants in maize
Chemical composition alteration: *e.g.*, low euricic acid mutants of rape seed
Male sterility: for use in hybrid breeding in various crops
Horticultural variants: development of various floral mutants
Breeding of asexually propagated species: numerous species and traits
Development of genetic stock: various lines for breeding and research
Development of earliness: achieved in many species

b) Induced Mutants

☆ **Rice:** Jagannath-gamma ray induced mutant from T.141 (by H. K. Mohanthy and S. S. Mishra at CRRI, Cuttack)

☆ **Wheat:** Sarbati Sonora Gamma radiation from Sonora 64 (by M. S. Swaminathan and Vergees), NP 836 mutants, through irradiation from NP 709 (by M. S. Swaminathan) and Pusa lerma.

☆ **Cotton:** Indore 2 -Induced from Malwa upland 4; MLU 7 - gamma ray induced mutant from culture 1143 EE MLU 10 -gamma ray induced mutant from MLU 4.

☆ **Mustard :** Primax whicte (1950); Summer Pope Seed Regina I (1953)

☆ **Sugarcane :** Co 8152- gamma ray induced mutant from Co 527

☆ **Groundnut :** NC 4, TG 3, TG 1 (Vikram) in 1973 by S. H. Patel

☆ **Castor :** Aruna (NPH1) – Fast neutrons induced mutant from HC 6

☆ **Mungbeen :** Pant moong 2 in 1983 by D. D. Singh, TAP 7 in 1983 by S. E. Pawar

☆ **Black gram:** TAU 1 and TPV 4 by S. E. Pawar

- ☆ **Chickpea:** Pusa 408 (Ajay), Pusa 413 (Atul).
- ☆ **Sorghum:** SPV 1 to 6, Co 21 (striga resistance)

Table 40: Total Varieties Released through Mutation Breeding

Crops	*Total Varieties Released*
Field crops	
Cereals	1072
Pulses	311
Commercial	81
Vegetables	66
Oil seeds	59
Others	111
	1700
Ornamentals	**552**
Total	**2252 (from 175 crops)**

Source: IAEA-2000.

18

Polyploidy and Crop Improvement

Terminologies

Haploid: An individual with a gametic chromosome number (n).

Diploid: An organism having two copies of a single genome *i.e.,* with chromosome number of 2x.

Genome: It consists of all the chromosomes of a diploid species that are distinct from each other with respect to their gene content and, often morphology.

Monopliod: An individual with the basic chromosome number (x) *i.e.,* with one genome.

Polyploids: Individuals having more than two identical or distinct genomes.

Table 41: Classification of Polyploidy

Ploidy	*Genome*	*Description*
Diploidy	AA BB	Contains two of a basic chromosome set
Euploidy		
Autoploidy	AAA BBBB	Multiples of a basic set(n) of one specific genome
Alloploidy	AAB AABB	Multiples of the basic number but of different genomes
Segmental alloploidy	AA'B AB'B'	Multiples of the basic number but the genomes have similar parts
Aneuploidy	AA	2n + or -1,2,......, k

AUTO POLYPLOIDY

Individuals having similar genome represented more than twice.

e.g. Auto triploid 3 x, Auto tetraploid 4 x, Auto pentaploid 5 x, Auto hexaploid 6 x, *etc.*

Origin and Production of Doubled Chromosome Numbers

1. **Spontaneous**: Chromosome doubling occurs occasionally in somatic tissues and unreduced gametes are also produced in low frequencies. Production of unreduced gametes are promoted by certain genes *e.g.* genes causing asynapsis and desynapsis.
2. **Production of Adventitious buds**: Decapitation in some plants leads to callus development at the cut end of the stem. Such a callus has some polyploid cells, and some shoot buds regenerated from the callus may be polyploids. *e.g.* Solanaceae
3. **Physical agents**: Heat or cold treatments, centrifugation and X-ray and gamma ray irradiation will produce polyploids. Tetraploid branches were produced in Datura by cold treatment. Heat treatment is successfully used in barley and wheat.
4. **Regeneration *in vitro***: Plants of various ploidy have been regenerated from callus cultures of *Nicotiana*, Datura, Rice.
5. **Colchicine treatment**: It is the most effective and most widely used treatment for chromosome doubling. Both dicot and monocot plants are successfully utilised for colchicine treatment. The chemical arrests the formation of spindle fibre resulting in stoppage of movement of chromosomes to opposite poles.

Morphological Features of Auto Polyploids

1. They have larger cell size than diploids.
2. Pollen grains are generally larger.
3. Polyploids are slower in growth and late in flowering.
4. Polyploids have larger and thicker leaves, larger flowers and fruits which are usually lesser in size.
5. They show reduced fertility.
6. Different species have different levels of optimum ploidy. *e. g.* Sugar beet - Triploid.
7. They have lower dry matter content.

Application of Autopolyploidy in Crop Improvement

Triploids

- **Triploid Watermelon:** Seedless watermelons are grown commercially in Japan. They are produced by crossing tetraploids (4x), used as female with diploids (2x) as male. The reciprocal cross is not successful.
- **Triploid Sugar beet**: Have larger roots and more sugar per unit area than diploids and tetraploids. 3x is the optimum level of ploidy in sugar beets. Seed production is done by planting 4x and 2x plants.
- **Triploid banana:** Is another example for use of triploid. Once triploids are produced they are propagated vegetatively.

Auto Tetraploids

- **Tetraploid Maize**: Have 43 per cent more carotenoid pigments and Vitamin A than diploid. In forage crops tetraploid red clover, rye grass, berseems are successful. They are more vigorous, digestible and palatable. In ornamental crops, tetraploids are successful, because of increased flower size and longer flowering period.

Limitations

1. Autopolyploids are successful in species with lower chromosome number.
2. Cross-pollinatedspecies are more responsive than self-pollinating species.
3. Larger size of autopolyploids is generally accompanied with higher water content.

This is not a desirable character in cabbage and turnip.

4. In crops grown for seed exhibit high amount of sterility though seed size is increased.
5. Triploids cannot be maintained except through clonal propagation.
6. New autopolyploids cannot be used directly as crops because they will have some undesirable characters *e.g.* poor strength of stem in grapes, irregular fruit size in water melon.
7. Effects of autopolyploidy cannot be predicted.

Allopolyploidy

Individuals having more than two genomes which are dissimilar. Several of our crop plants are allopolyploids. *e.g.* Tetraploid Cotton, Tetraploid Ground nut, Tetraploid Wheat, Tetraploid Ragi, Tobacco, *etc.*

Origin and Production of Allopolyploids

The present day allopolyploids were most likely produced by chromosome doubling in F_1 hybrids between two distinct species. This chromosome doubling might have occurred in somatic tissues due to irregular mitotic cell division or due to irregular meiosis leading to unreduced gametes.

Role of Allopolyploids in Evolution

Allopolyploids are more successful as crop species than autopolyploids. Many of our present day Crop species are allopolyploids. Allopolyploid has contributed to a greater extent in evolution of plants. This is evident from wide spread occurrence of allopolyploids in various genera of plants. *e.g. Brassica, Gossypium, Triticum, Eleusine, Avena, Arachis, Nicotiana, Saccharum, Solanum, etc.*

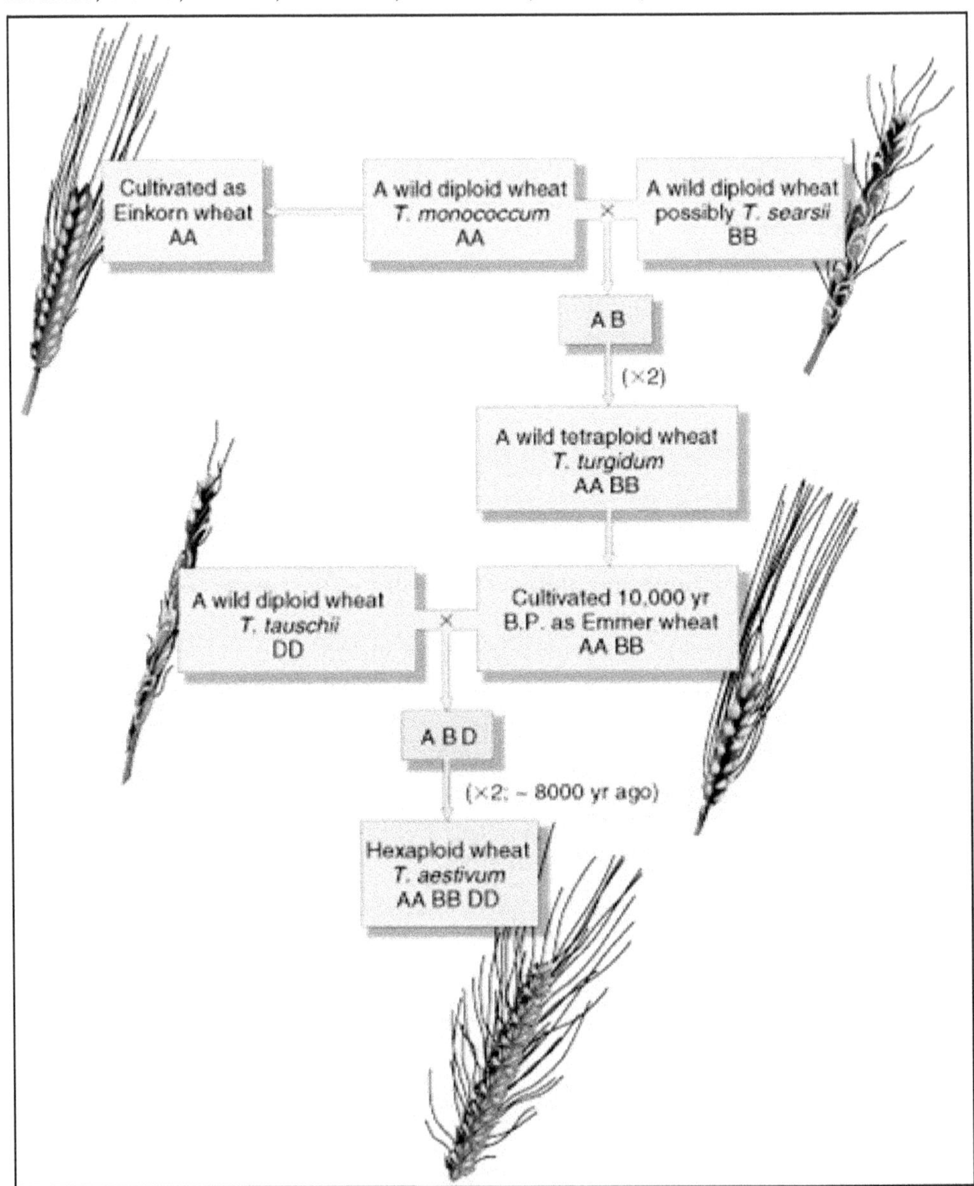

Figure 37: The Proposed Origin of Common Wheat *Triticum aestivum*.

It is estimated that one third of angiosperms are polyploids and in that majority of them are allopolyploids. It has been possible to trace back the evolutionary history of many allopolyploid crop species and diploid crop species and diploid parental species.

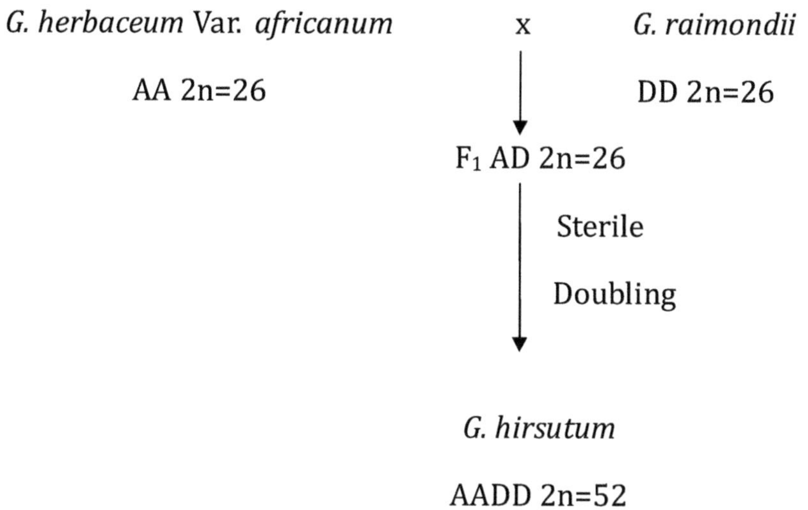

Figure 38: The *G. hirsustum* Evolutionary Pathway.

Application of Allopolyploids in Crop Improvement

1. **Utilization as a bridging species:** Amphidiploids serve as a bridge to transfer a character from one species to related species, generally from a wild species to a cultivated species.

   e.g. Use of *N. digluta* for transfer of resistance to tobacco mosaic virus from *N. sylvestris* to *N. tabacum*.

2. **Creation of new crop species:** *e.g. Triticale, Raphano brassicas, etc.*
3. **Widening the genetic base of existing allopolyploids:** The genetic base of *B. napus* is narrow. To widen the genetic base it has been crossed to *B. campestris* and *B. oleracea*

Limitations of Allopolyploids

1. The effects of synthesized allopolyploids cannot be predicted. *e.g. Raphano brassicas*
2. Newly synthesized alloploids have many defects. *e.g:* Low fertility, unstable and other undesirable features.
3. The production of useful synthetic allopolyploids require much time and labour.

Aneuploids

Classification

- **Monosomic**: One chromosome missing 2n-1. *e.g.* Tobacco, Wheat, *etc.*
- **Nullisomic**: One chromosome pair missing. 2n-2. *e.g.* Wheat and Oats.
- **Double Monosomic**: One chromosome from each of two different chromosome pairs missing. 2n-1-1.
- **Trisomic**: One chromosome extra 2n + 1. *e.g.* Datura, Maize, *etc.*
- **Double Trisomic**: One chromosome from each or two different chromosome pairs extra

  2n + 1 + 1.
- **Tetrasomic**: One chromosome pair extra 2n + 2.

Origin of Aneuploids

1. **Spontaneous**: Due to meiotic irregularities there may be occurrence of monosomics (2n - 1) and Trisomics (2n + 1). *e.g.* Datura
2. **Triploid Plants**: Distribution of chromosomes at the meiosis I Anaphase is irregular leading to the production of whole range of aneuploids in the progenies.
3. **Asynaptic and Desynaptic plants**: In these plants few to all chromosomes are present as univalent's at metaphase I of meiosis. In the progenies of such plants a relatively high frequency of aneuploids occurs.
4. **Translocation Heterozygotes**: The translocation heterozygote would produce n + 1 and n –1 gametes. As a result variable frequencies of aneuploids are found.
5. **Tetrasomic Plants (2n + 2):** They produce n+1 gametes. When they are crossed with normal diploid plants, they produce a high frequency of trisomic plants (2n+1).

Applications in Crop Improvement

1. Aneuploids are useful in studying the effects of loss or gain of an entire chromosome or chromosome arm on the phenotype of the individual.
2. Aneuploids are useful in locating linkage group on to a particular chromosome. By using a secondary or tertiary trisomic, the gene may be located to one of the two arms of a chromosome or even to a part of the chromosome arm.
3. Aneuploids are useful in identifying the chromosomes involved in translocation.
4. They are useful in the production of substitution lines.

Limitations of Aneuploids

Production, identification and maintenance of aneuploids require elaborate cytogenetic analysis, which is difficult, time consuming and requires considerable skill.

19

Ideotype Concepts in Crop Improvement

- ☆ The term Ideotype breeding was introduced by "Donald" 1968.
- ☆ **"Ideotype"** is a biological model which is expected to behave in a predictable manner with in a defined environment".
- ☆ **Crop Ideotype** is a plant model which is expected to yield a greater quantity or quality of grain, oil, other useful product when developed as a cultivar.
- ☆ The Ideotype has following synonyms- model plant type/Ideal model plant type/Ideal plant type.

Steps in Ideotype Development

1. Definition of the target area and target environment.
2. Quality consideration.
3. Current agronomic practices.
4. Draw a list of desired as model plant type.
5. Assessment of traits in target environment.
6. Simulation models have been used as aids to trait analysis.
7. The choice of traits to be included in an Ideotype will also depend up on certain consideration.
   - ☆ The trait should enhance yield.
   - ☆ Ease and speed in measurement of trait.
   - ☆ Reasonable stability in the expression of traits.

- ☆ Heritable variations must be present for the trait.
- ☆ Measurement of trait should be in expensive and reliable.
- ☆ High heritability of the traits.

Characteristics of a Crop Ideotype

1. It should be a weak competitor.
2. It should efficiently utilize environment resource.
3. Ideotype is a moving goal.
4. It is sensitive to cultural practices.
5. It should include morphological and physiological characteristics that result in high harvest index.
6. It should be grown as far as possible in weed free situations.
7. A conscious and planned selection has to be done for the selected crops.

Ideotype of Selected Crops

The wheat ideotype defined by Donald (1968) comprised the following:

1. A relatively short and strong stem.
2. A single culm.
3. Erect leaves (near-vertical).
4. A large ear.
5. An erect ear.
6. Simple awns *etc.*

Important Ideotype Concepts

- ☆ Barley – Donald (1979)
- ☆ Six rowed barley – Rasmussen (1987)
- ☆ Maize – Mock and Pearce (1975)
- ☆ Rice – Matsushima (1964)
- ☆ Concept of Plant Type - Jennings (1964)
- ☆ Dwarf rice – chandler (1969)
- ☆ Cotton – Singh and Narayan
- ☆ Brassica – Bhargare (1984).

Ideotype Breeding

"Ideotype breeding aims to enhance genetic yield potential by modifying individual traits to their pre-defined optimum levels.

Seeps in Ideotype Breeding

1. Defining the phenotypic goals.
2. Creation of adequate genetic diversity.
3. Selection of lines having desirable phenotype.
4. Continued breeding efforts

   (Evaluation of phenotypes in several genetic and cultural back ground)

Factor that Slow Breeding Process

1. Symmetry in size of plant parts
2. Compensation among plant parts
3. Pleiotrophy
4. Inferior Genetic background

Merits of Ideotype Breeding

1. Improvement in yield.
2. Creation and maintenance of genetic diversity.
3. Primary gene pool of a crop may be divided into improved and unimproved gene pools.
4. It encourages development and evaluation of hypothesis regarding how yield is achieved.
5. The goals for specific traits are established with a view to maximize their effects in yield.

Table 42: Difference between Traditional and Ideotype Breeding

Sl.No.	*Traditional Breeding*	*Ideotype Breeding*
1.	The breeder has an idea to develop type but he does not describe it formally.	The breeding must define Ideotype to be developed based a trait analysis and other considerations.
2.	Selection uses yield per se as a criterion	Yield is not used as a basis of selection
3.	It is usually based an improved gene pool of a crop.	It will usually introgress desirable genes or traits from unimproved gene pool.
4.	It does not encourage systematic thinking and accumulation of information on how yield as achieved in a crop.	It dose
5.	It dose not deliberately generate genetic diversity.	It generates genetic diversity deliberately
6.	Progress is relatively rapid	Progress is slow
7.	It constitutes the main activity and is likely to remain so.	It may be considered as augment traditional breeding effort and not to replace it

Limitations of Ideotype Breeding

1. It has not been possible to identify individual traits with enhanced yield universally or relatively limited genetic and environmental situations.
2. Use of physiological and biochemical traits in the Ideotype concept is cumbersome, costly and time taking.
3. New methodologies including techniques and instruments need to be developed so that breeder can include a large number of traits.
4. Transfer of traits from unimportant gene pool would often be necessary; this would involve considerable breeding efforts.
5. Development of single plant type may narrow the genetic diversity within the crop leading to increased genetic vulnerability.
6. Ideotype breeding has been sometimes accused of being pre-occupied with extreme morphological characters.

20

Breeding for Resistance to Abiotic Stresses

Importance of Abiotic Stresses

Only about 30 per cent of the earth is land. Of this, about 50 per cent is not suitable for economic crop production, mainly because of constraints of temperature, moisture and topography. Of the remaining portion of arable land, optimum production is further limited by a variety of environmental stresses, requiring mineral and moisture supplementation for economic crop production. As the world population increases more food will have to be produced by increasing the productivity of existing farmlands as well as bringing new lands into production. This means, marginal lands will have to be considered. Plants breeders will have to develop cultivars that are adapted to specific environmental stresses. It is estimated that abiotic environmental stresses are responsible for about 70 per cent of yield reduction of crops in production.

Characteristics of Abiotic Stresses

1. The characteristics of an abiotic stress may vary considerably depending on the location.
2. The occurrence and the degree of some of the stresses are unpredictable.
3. The degree of some stresses is likely to vary during the crop season.
4. Some stresses can be relieved by appropriate management practices (*e.g.* drought, salinity *etc.*), while others are virtually impossible to manage (*e.g.* temperature stresses).

5. Different plant/crop species show difference in their abilities to withstand a given stress.
6. Different growth stages of crop may show marked difference in their tolerance to an abiotic stresses.
7. Stress during the reproductive phase of crop causes far more economic loss than comparable stress during the earlier phases.
8. The effects generated by one abiotic stress may overlap some of those generated by another stress.

Types of Abiotic Environmental Stresses

1. **Drought**: This is the environmental condition caused by lack of rainfall.
2. **Heat**: Heat stress occurs when temperatures are high enough to cause irreversible damage to plant function.
3. **Cold**: Cold stress manifests when plants are exposed to low temperatures that cause physiological disruptions that may be irreversible.
4. **Salinity**: Stress from salinity occurs when the dissolved salts accumulate in the soil solution to an extent that plant growth is inhibited.
5. **Mineral toxicity**: Mineral toxicity occurs when an element in the soil solution is present at a concentration such that plants are physiologically impaired.
6. **Oxidation**: Oxygen-free radicals (or activated oxygen) are known to cause degenerative conditions in plant cells.
7. **Water-logging**: Excessive soil moisture as a result of prolonged rainfall can cause anoxic soil conditions, leading to roots suffering from lack of oxygen.
8. **Mineral deficiency**: Inadequate amounts of essential soil minerals for plant growth cause crop injury.

Minimising Losses due to Abiotic Stresses

I. **Crop Management:** The degree of stress could be reduced by suitable agronomic/soil management practices. *e.g.*: irrigation to relieve drought stress, soil amendments to reduce salinity.

II. **Development of Resistance Variety(s):** This offers the cheapest, easiest to apply and most eco-friendly approach to reduce losses due to any stress.

   a. **Indirect:** Evaluation of materials under stress often helps in identification of a resistant line that may become an important commercial variety.

   b. **Direct Breeding:** Deliberate breeding of materials for stress resistance. When a genetically variable material is grown under the stress environment, selection can be based on survival, yield and other traits contributing to stress resistance.

BREEDING FOR DROUGHT RESISTANCE

Water is the most limiting factor in crop production. In tropical regions of the world, moisture extremes are prevalent. There is either too much of it when the rain falls, or there is little or lack of rainfall. Drought is responsible for severe food shortages and famine in developing countries.

Definition

"The inadequacy of H_2O availability including precipitation and soil moisture storage capacity in quantity and distribution during the life cycle of a crop to restrict explanation of its full genetic yield potential".

Drought Resistance Mechanisms

The various mechanisms by which a crop can minimize the loss in yield due to drought are grouped in the following four categories.

1. **Drought Escape:** Using early maturing cultivars may allow the crop to complete its life cycle (or at least the critical growth stage) before the onset of drought later in the season. The plants use the optimal conditions at the beginning of the season to develop vigour.
2. **Dehydration Avoidance:** Some plants avoid drought stress by decreasing water loss, for example by having cuticular wax or by having the capacity to extract soil moisture efficiently.
   a. **Reduced Transpiration**
      - **Stomatal sensitivity**: Closure of their stomata in response to water deficit well before wilting.
      - **Osmatic adjustment**: As it allows growth and results in delayed leaf death beyond a given threshold would not reduce transpiration.
      - **Cuticular wax**: Helps in reducing the water loss due to transpiration.
      - **Abscisic acid:** Decrease in leaf expansion and helps in promotion of root growth.
      - **Leaf pubescence:** Increase the leaf reflectance and reduce the leaf temperature.
      - **Leaf angle and leaf movement:** Erect leaves prefer over lax leaf lines.

   b. **Increased H_2O uptake:** Deep root system, root length density, root hydraulic resistance and increase in root growth are the important traits contribute to the maintenance of higher water potential.
3. **Dehydration tolerance:** In species such as cereals in which grain filling is found to depend on actual photosynthesis during the stage, as well as dry matter distribution from carbohydrates in pre-anthesis, terminal drought significantly reduces photosynthesis. This shifts the burden of

grain filling to stored carbohydrate as the source of dry matter for the purpose. Consequently, such species may be more tolerant of post anthesis drought, being able to produce appreciable yield under the stress.

- Maintenance of membrane integrity: It determined by the leakage of solutes from cells.
- Seedling survival
- Seedling growth
- Seed germination under osmotic stress
- Presence of large amounts of awns.
- Proline accumulation.

4. **Recovery**: Because drought varies in duration, some species are able to rebound (recover) after a brief drought episode. Traits that enhance recovery from drought include vegetative vigor, tillering, and long growth duration.

Genetic of Drought Resistance

1. **Oligogenic**
   - Waxy bloom, Glossy trait, Glaucousness, Glabrous leaves
   - ABA accumulation in Wheat, Proline accumulation in Barley, resistant to flower dropping and pod formation in Rajma.
2. **Polygenic:** Many traits

Sources of Drought Tolerance

- ☆ *Cultivar*
- ☆ *Landraces*
- ☆ *Wild relatives*
- ☆ *Transgenes.*

Table 43: Some Successful Examples of Transgenes Used for Abiotic Stress Tolerance

Transgene	*Isolation*	*Function*	*Transferred into*	*Protection from*
Hsf	–	Heat shock factor	Tobacco	Heat stress
P5CS	Mothbean	Pyrrolines 5 carboxylate Synthetase	Tobacco	Salt stress
mtl 1D	*E. coli*	Mamitol-1-phosphate Dehynogenase	Tobacco	Salt stress
hva1	Barley	A class 3 lea protein	Rice	Salt stress
Sac B	*B. subtilis*	Fractan accumulation	Tobacco	Drought
Sal 1	Arabidopsis species	Sulphur assimilation	Yeast	High salt stress

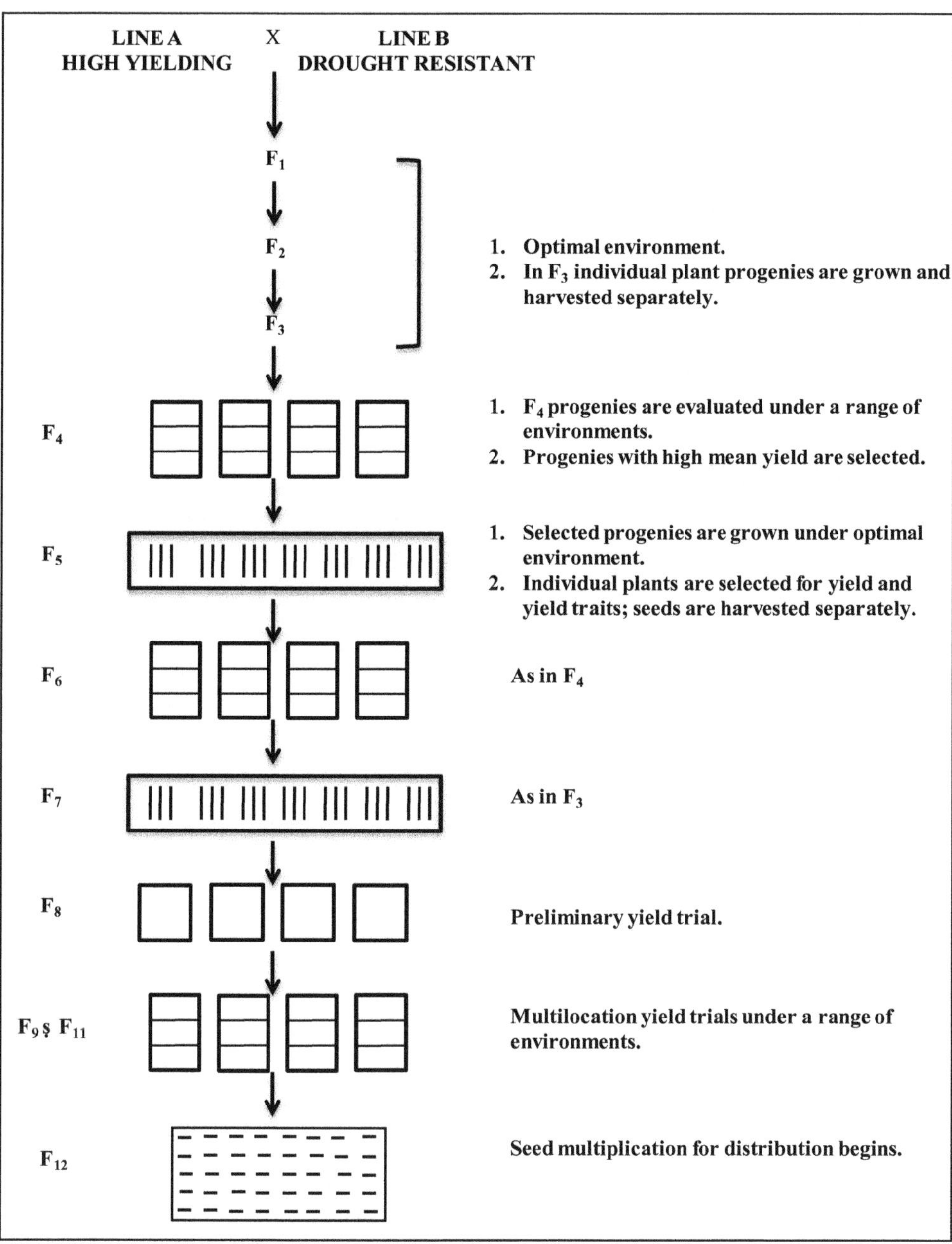

Figure 39: A Breeding Approch to Develop Varieties Suitable for a Range of Environments.

Selection Criteria

The selected trait should have the following features.

☆ It should be easy to estimate/score.

- ✰ Highly heritable.
- ✰ Large genetic variability should exist for the trait.
- ✰ Significant associate with drought resistance.
- ✰ Positive association with yield under stress.

Breeding Approaches

1. Adaption to a specific environment : Line A (High yields) x Line B (Drought resistance).
2. Variable environment:
3. Combining selection for drought resistance traits and high yield potential.
4. Shuttle breeding in wheat: CIMMYT, Mexico.

Difficulties in Breeding

1. Variation in moisture regime year to year and location to location.
2. Selection for drought resistance is difficult.
3. Selection indices demands considerable resources, effort and attention.
4. Measurement of many drought resistance traits is different and problematic.
5. Many drought resistance traits may reduce yield.
6. Wild relatives – problematic and their value in improving are questionable.

BREEDING FOR SALINITY RESISTANCE

Soil salinity constraints to crop production occur in an estimated 95 per cent million hectares worldwide. Such soils are mainly of two types: (1) Saline and (2) alkaline soils. In addition, (3) acidic soils forms the third group and are characterized by mineral toxicities/deficiencies.

Salinity

Is the accumulation of dissolved salts in the soil solution to a degree that inhibits plant growth and development.

Management of Salt affected Soils

1. **Reclamation:** Removal from root zones (salts/exchangeable sodium) by leaching (salts) application of suitable Ca^{++} amendments to remove excess exchangeable Na^{+} salts and suitable drainage.
2. **Appropriate management:** Variety, method and time of planting, irrigation and other cultivation practices.

   e.g.: Crop tolerant to saline/alkali soils:

   Rice: Arya33, Bhura Rata, damodar, Pokkali, AV-1, CSR-1, *etc.*

   Wheat: Rata, Kharchia, *etc.*

Crops different their Tolerant to Salinity

- Sensitive: Apple, carrot, bean, onion okra, orange
- Moderately Susceptible: Alfatfa, cabbage, maize, cowpea, potato, sugarcane, tomato
- Moderately Tolerant: Barley, broccoli, safflower, sorghum, soybean, wheat
- Tolerant: Barley (grain), cotton, sugarbeet.

Estimation of Salinity Resistance and Selection Criteria

1. Cell survival
2. Seed germination
3. Dry matter content
4. Leaf death and senescence
5. Leaf in content
6. Leaf necrosis
7. Root growth
8. Osmoregulation
9. Yield

Sources of Salinity Resistance

1. Cultivated varieties
2. Germplasm collection – Kharchia, wheat, Rajasthan
3. Related sps – *Agropyron*>wheat, *L. cheesmanii*>tomato
4. Somaclones
5. Transgenes

Breeding Approaches

1. Selection
2. Hybridization: Kharchia 65 x WL711 – KRL1-4 -20- 30q/ha under 10-15dsm^{-1}
3. Inter-specific hybridization: *L. esculentonn* x *L. Cheesmanii* variety minor Walter – germinate less than 300 mm NaCl
4. Cell selection
5. Genetic engineering

Limitations

1. Selection work is tedious, costly and beyond reach of many breeders

2. No simple to score, reliable and dependable selection criterion for salinity resistant
3. Genetic control is complex and polygenic, which makes transfer difficult
4. Basis for salinity resistant is poorly understood

21

Breeding for Resistance to Biotic Stresses (Disease and Insect Pest)

What is Disease?

- ☆ Abnormal condition in the plant produced by an organism.
- ☆ "Physiological disorder or structural abnormality that is deleterious to the plant or to any of its part or products, or that reduces their economic value" (Stakman and Harrar, 1957).
- ☆ "Harmful deviation from normal functioning of physiological processes" (Anon, 1950).

Different Pathogens causing Diseases

Different pathogens causing diseases showing in Figure 40.

Losses due to Diseases

Disease reduces biomass (dry matter) and hence yields. This may happen in either of the following ways:

- ☆ **Killing of plants** (*e.g.* vascular wilts, various soil borne fungi), general stunting caused by metabolic disruption, nutrient drain or root damage (*e.g.* many viruses).
- ☆ **Killing of branches** (*e.g.* some fungal diebacks).
- ☆ **Damage to leaf tissues** (*e.g.* many rusts, mildews, blights, leaf spots).

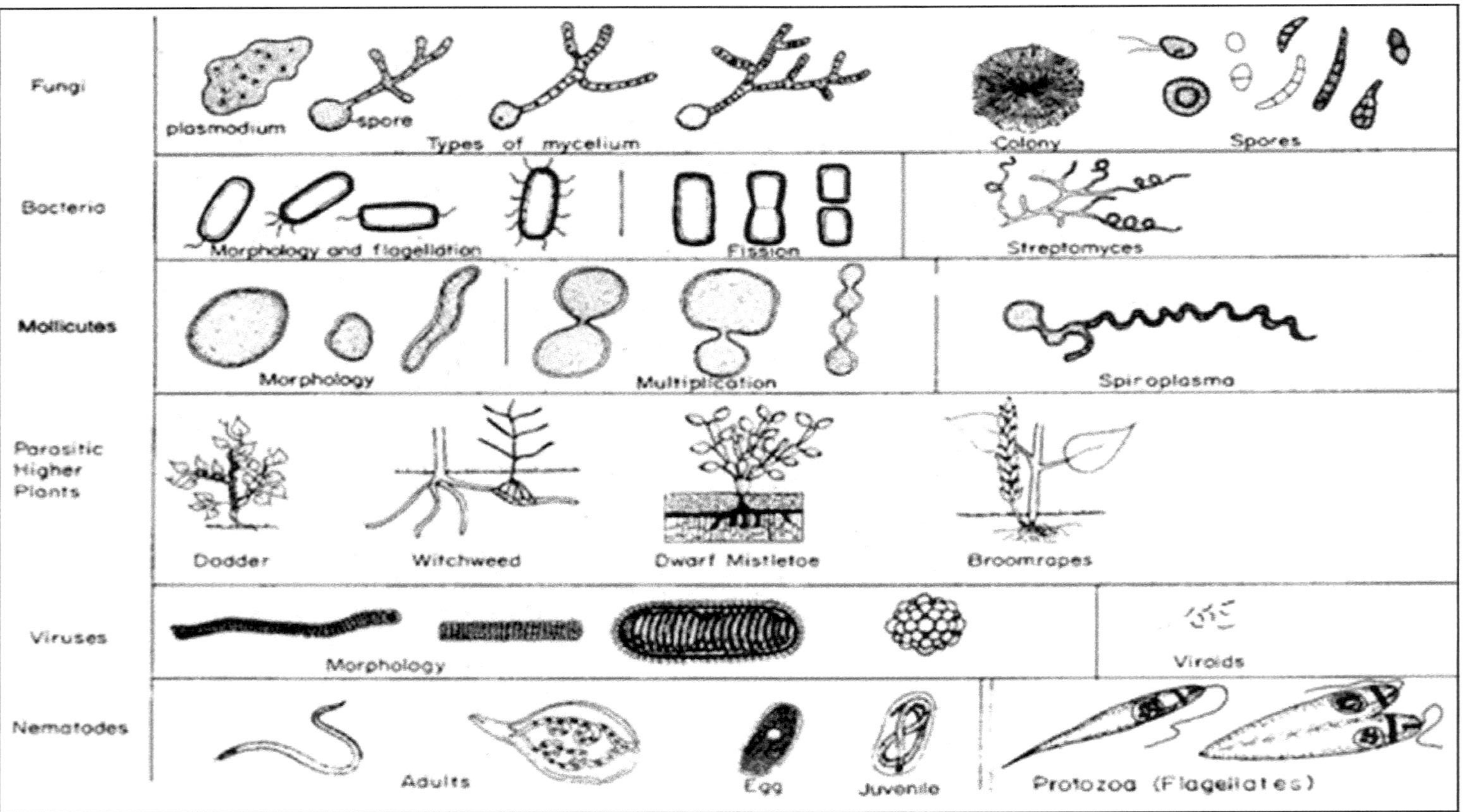

Figure 40: Morphology and Multiplication of some of the Groups of Plant Pathogens.

- **Damage to reproductive organs including fruits and seeds.**

Often the loss due to diseases may range from a few to 20 or 30 per cent; in case of severe infection, the total crop may be lost (Jayaraj, 2002).

MAJOR FAMINES OR FOOD LOSSES

- **Irish famine** (**1840's**) due to **potato leaf blight** epidemics;
- The **wheat less days (1917's)** in **USA**, due to **stem rust** epidemics;
- Devastation of all **Victoria-derived oats (mid 1940's)** in **USA** due to fungus causing **Victoria blight** disease;
- **Bengal famine** (**1943's**) due to **brown spot** (*Drechslera oryzae)* disease of **rice**;
- **Downy mildew** epidemic (**1973's**) in **pearl millet** (c.o. *Sclerospora graminicola*) in **India**
- **Bacterial blight** (*X. campestris* pv. *oryzae*) (**1980's**) severe out-break in **Panjab** (A. P. K. Reddy 1980).

History of Breeding for Disease Resistance

- **Theophrastus** (3rd Century): cultivated variety differed in their ability to avoid diseases.
- **Benedict Prevost**: diseases are produced by pathogen. He showed that wheat bunt was produced by a fungus.
- **Andrew Knight** (Mid 19th Century): crop varieties differed for disease resistance.
- **Biffen** (1905)**:** resistance to yellow rust in wheat was governed by a recessive gene segregating in the ratio 3:1 in F_2.
- **Erikson** (1894): Pathogens, although morphologically similar, differed from each other in their ability to attack different related host species.
- **Barrus** (1911): different isolates of a microorganism differed in their ability to attack different varieties of the same host species; this finding is the basis for physiological races and/or pathotypes.
- **Flor** (1956): Gene-for-Gene Hypothesis.

Disease Development

Disease development showing in Figure 41.

Stages of Disease Development (especially fungal diseases)

- **Contact**: landing of pathogen on host tissue.
- **Infection**: Pathogen gains entry in to the host tissue.
- **Establishment**: Pathogen proliferates and spreads within host tissue.
- **Development**: Spore production/multiplication by pathogen and symptoms developed.

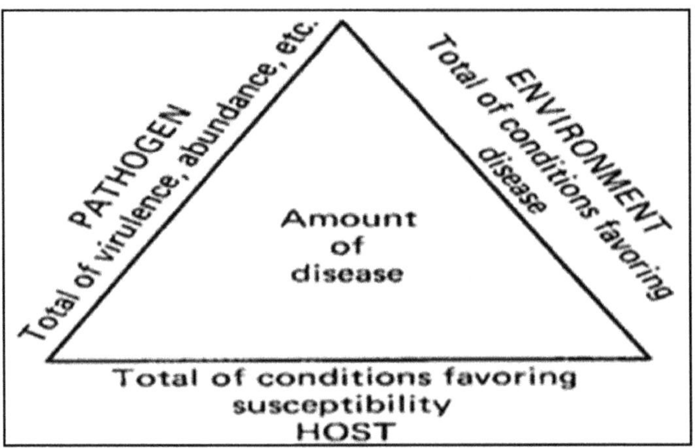

Figure 41: The Disease Triangle.

Disease Reactions

Various reactions of hosts to the pathogen may be grouped into;

- ☆ **Susceptible**: Disease development is profuse and is presumably not checked by host genotype. *e.g.* Agra Local susceptible to wheat rust (r=1).
- ☆ **Immune**: Host does not show symptoms of a disease (100 per cent freedom from infection, r=0).
- ☆ **Resistance**: Less disease than susceptible (r>0 and r<1).
- ☆ **Tolerant**: A tolerant cultivar is one which endures disease attack (R. M. Caldwell *et al.* 1958) and looks susceptible one (Browning and Frey 1969).

Management of Resistant Genes

Vertical Resistance: Van der Plank (1963, 1968)

- ☆ Specific resistance of host to particular race of a pathogen.
- ☆ Also called oligogenic, race specific, pathotype specific or specific resistance.
- ☆ Controlled by one or few major genes.
- ☆ When controlled by single major gene it is known as monogenic resistance.
- ☆ Each gene has large and easily identifiable effect on resistance.
- ☆ Less stable as can be overcome by appearance of one virulent gene.
- ☆ VR involves hypersensitive reaction to the pathogen.
- ☆ Less effect of environment.
- ☆ VR governed by two or more major gene is more durable than controlled by single major gene.
- ☆ Very effective against few races but totally ineffective against others.
- ☆ Transfer from one host to other is simple.

Horizontal Resistance

- ☆ Resistance of host to all the races of pathogen.
- ☆ Reproduction rate of pathogen is never zero, but is less than one, (*i.e.* r>0 but <1).
- ☆ Also called as race non specific, pathotype non specific, polygenic, partial or general resistance.
- ☆ Controlled by polygenes.
- ☆ Provides protection from all races of a pathogen.
- ☆ It has continuous variation and hence, impossible to classify plants into different clear-cut classes.
- ☆ More durable than VR since it involves several features (polygenes) of host plant.
- ☆ More stable since it require new genes for virulent that is less probable.
- ☆ Influenced by environmental factors.
- ☆ Transfer of HR from one host to another is more difficult than VR.

Mechanism of Disease Resistance

Mechanical

- ☆ Certain **mechanical and/or anatomical features** of host may prevent infection.

  e.g. closed flowering habit of wheat and barley prevents infection by spores of ovary infecting fungi (Macor, 1960).
- ☆ Several leaf characters of rice cultivars *viz.* rolled, narrow and dark green were associated with resistance to bacterial leaf blight (X. *campestris* pv. *oryzae*) resistance (C. B. Singh and Y. P. Rao, 1971).

Hypersensitivity

- ☆ Immediately after infection several host cells surrounding the point of infection die leading to death of pathogen or prevents spore production.
- ☆ Found in biotrophic or obligate parasite (*e.g.* C.O. of rust, smut *etc.*).
- ☆ Hypersensitive cell produce chemical/phenolic compounds (phytoalexins, *etc*) which are fungitoxic and autotoxic (Deverall 1976).
- ☆ In large no. of cases, immune reaction is due to hypersensitive reaction of host.

Nutrional

- ☆ The reduction in growth and in spore production is generally supposed to be due to unfavourable physiological conditions within the host.
- ☆ Most likely, a resistant host does not fulfill the nutritional requirements of pathogen and hence limits its growth and reproduction.

Table 44: Genetics of Disease Resistance in some Crop Plants

Genetic Control of Inheritance	*Crop Species*	*Disease*	*Expression of resistance*
Oligogenic	Wheat	Yellow rust	1D, 2D, 3D
		Stem rust	2D
		Leaf Rust	1D + m
	Rice	Blast	1D
		Rungro virus	1D, 3D +1I
		Bacterial blight	2D
	Maize	Southern corn	1D
		Corn disease	1D + 1d
	Sorghum	Smut	3D
		Milo disease	1d
	Cotton	Fusarium wilt	1D.2D
		Verticillium wilt	1D + m.1d
	Tomato	Fusarium wilt	1D + m
		Leaf mould	1D
		Tomato mosaic	2D
Polygenic	Maize	Maize rust	Polygenic
		Southern corn	
		Leaf blight	
	Rice	Bacterial blight	
	Cotton	Bacterial blight	
		Reniform nematode	
	Tomato	Bacterial canker	
		Bacterial blight	
		Curly top virus	
Cytoplasmic	Maize	Southern corn	Cytoplasmic
		Leaf blight	
	Wheat	Leaf rust	

Gene-for-Gene Concept

- ☆ Postulated by Flor in 1956 based on his work on Linseed rust (c.o. *Melampsora lini*).
- ☆ For each resistance gene (*R* gene) in the host, there are corresponding avirulence genes (*avr* gene) in the pathogen (Flor, 1956).
- ☆ Only one of the four combinations would lead to the resistant response since the products of *R* and *A* would recognize and interact with each other.

- ☆ The product of alleles *a* and *r* are unable to recognize each other, and there is no interaction between them hence reaction of host becomes susceptible.

Resistance Gene	*Virulence Gene*	
	A	*a*
R	Resistance	Susceptible
r	Susceptible	Susceptible

- ☆ Allele *A* of the virulence gene specifies **avirulence.**
- ☆ Allele **a** of the virulence gene governs **virulence**

Sources of Disease Resistance

- ☆ A known variety
- ☆ Germplasm collection
- ☆ Related species
- ☆ Mutants
- ☆ Somaclonal variants
- ☆ Unrelated organisms

Methods of Breeding for Disease Resistance

Commonly used breeding methods are as follows.

Selection

- ☆ Cheapest and quickest method.
- ☆ The resistant plants may be multiplied, screened and released as a variety.
- ☆ **Kufri Red,** potato from Darjeeling Red Round.
- ☆ **Pusa sawani,** okra (*A. esculentus*) from Bihar is resistant to yellow mosaic under field condition.
- ☆ **MCU1,** Cotton (*G. hirsutum*) from CO-4 for resistance to black arm.

Introduction

- ☆ The resistant variety may be introduced and after testing, if found suitable, can be released in the disease prone area.
- ☆ Easy and quick
- ☆ Introduction also serves as source of resistance in breeding programmes.

Mutations

- ☆ Induced mutations can be utilized by direct release as a variety (resistant mutant) and by utilization in hybridization programme.

e.g. Co 8153, a sugarcane variety developed from variety Co 775 with the use of gamma-ray irradiation was resistant to smut.

Hybridization

Used for:

- ☆ Transfer of disease resistance from an agronomically undesirable variety to a susceptible but otherwise desirable variety (by **backcross method**).
- ☆ Combining disease resistance and some other desirable characters of one variety with the superior characteristics of another variety (by **pedigree method**).

Marker Aided Selection

For ease in effective selection and accelerating the breeding programme morphological or biochemical or DNA markers will be used.

Somaclonal Variation

The variability generated by the use of a tissue culture cycle.

Somatic cell hybridization

Involves physical union or fusion of protoplast from two parents.

Meristem tip culture (for virus free planting material)

Limmaset and Conuet (1972) observed that in virus infected plants, the virus concentration decreases towards the growing point.

Transgenics for Disease Resistance

Transgenics for disease resistance showing in Tables 45(A) and (B).

Table 45(A): Virus Resistance Transgenics

Crop	*Transgene*	*Virus*	*Reference*
Tomato	CP-A1MV	A1MV	Tumer *et al.* (1987)
Tobacco	CP-PVX	PVX	Hemenway *et al.* (1988)
Potato	CP-PVY	PVY	Lawson *et al.*. (1990)
Tomato	CP-TMV	TMV, ToMV	Nelson *et al.* (1988)
Rice	Replicase protein	RYMV	Pinto *et al.* (1999)
Potato	Movement protein	PLRV, PVX, PVY	Tacke *et al.* (1996)
Potato	Antisense RNA	PLRV	Kawchuk *et al.* (1991)
Pea	Replicase protein	PSB MV	Jones *et al.* (1998)
Tomato	Ribozyme	CEVd	Atkins *et al.* (1995)
Potato	Antiviral protein ribonuclease	PSTV	Sano *et al.* (1997)

Disease Epidemics

- ☆ An epidemic is a severe outbreak of disease beginning from a low level of infection.
- ☆ Epidemics are common in case of air borne fungi.

Table 45(B): Transgenic Plants (Fungal and Bacterial Diseases)

Crop	*Transgene*	*Virus*	*Reference*
Tobacco	Bacterial chitinase	*Alternaria longipes*	Jones *et al.* (1988)
Tobacco	Bean chitinase gene	*Rhizoctonia solani*	Broglie *et al.* (1991)
Brassica napus	Bean chitinase gnene	*Rhizoctonia solani*	Broglie *et al.* (1991)
Tobacco	Barley X-thionin gene	*P. syvingae* *P.v.tabaci* *P. Syringae* *P.v. syringae*	Anzai *et al.* (1989)
Potato	Bacteriophase T-4 lysozyme	*E. casotovora sub sp. atroseptia*	During *et al.* (1993)
Potato	H_2O_2 gene for glucose oxidase	*V. daahalea,* *Phytophthora* *Erwinia carotovora*	Wu *et al.* (1995)
Tobacco	Stilbene synthase	*Botrytis cinerea*	Hin *et al.* (1996)

- ☆ An epidemic progresses from low infection level.

  e.g. An epidemic of potato leaf blight may be initiated by a single infector plant/sq. km.

Causes of Epidemics

- ☆ Narrow genetic base of cultivated crops
- ☆ Large acreages under single varieties
- ☆ Introduced pathogens
- ☆ Failure of VR genes (*e.g.* Vertifolia potato variety)
- ☆ Inadvertent breeding for susceptibility.

Measures for Prevention of Epidemics (Management of Disease Resistance)

Kush (1995) propose four approaches to prolong the useful life of a variety with VR

Deployment of Genes over Time

- ☆ Strong genes for VR may be employed or may be used to check epidemics in alternate **years** so that the virulent pathotype against any one of them doesn't evolve and doesn't survive even if it did evolve.

- ☆ Based on the principle of crop rotation to control certain soil borne pathogens.

Deployment of Genes over Space

- ☆ Strong genes for VR may be employed or may be used to check epidemics in **different geographical regions**; these regions would run perpendicular to the direction of movement of the pathogen.
- ☆ Thus each strong gene would act as a sieve for the pathotype virulent on the succeeding one(s).

Gene Pyramiding (Watson and Singh, 1953)

- ☆ Incorporation of diverse major genes for resistance to a pathogen/insect in one and the same variety.
- ☆ It is believed that, greater the number of major genes for resistance in a variety greater would be its longevity.

Limitations

- ☆ May lead to the evolution of "**super races**".
- ☆ If the super race in produced, which may overcome the combined resistance, then several resistance genes may be simultaneously rendered ineffective and may lead to quick exhaustion of available genes for resistance (Kush, 1995).

Multiline Variety

- ☆ One of the breeding approaches for a long-range control of diseases in SPCs is the development of multiline varieties.
- ☆ "A multiline variety is a population of plants that is agronomically uniform heterogenous for genes that condition reaction to a disease"
- ☆ This concept was first given by Jensen (1952) for use in oats.
- ☆ Similar approach was suggested by Borlaug (1959) in wheat for controlling stem rust.

Merits of Breeding for Disease

- ☆ Losses caused by disease crops are minimized with use of resistant varieties.
- ☆ Reduction in the cost of production resulting in increasing cost benefit ratio.
- ☆ Reduces environmental pollution and health hazards by reducing use of chemical molecules.
- ☆ They are non-toxic to man; farm animals and wild life (do not contain pesticidal resistance).

- ☆ Genetic resistance is only solution of some of the diseases such as wilts, rusts, smuts, nematodes and bacterial blights.

Demerits of Breeding for Disease

- ☆ It is long term process which takes 10-15 years to develop agronomically acceptable variety.
- ☆ In some cases, breeding for resistant to one disease leads to the susceptibility to another pests.
- ☆ Breeding for disease resistance is an expensive method (requires adequate finance for long period).

Table 46: Breeding Achievements for Resistance Breeding for Major Field Crops

Crop	*Major Diseases*	*Varieties (T/MT/MR/R)*
1. Paddy	Blast	Karjat 14-7(R), Palghar 60(R), Darna(R), R24(R)
	Bacterial leaf blight (BLB)	
2. Wheat	Stem rust, Black and Brown rust	Sonalika (S) HD2189 (MR), MACS 9(S), LOK 1 (S), Raj 1555(R), NIAW 34(R)
3. Sorghum	Head mould	Kharif: CS 3541(MT), SPV 351 (MT)
	Charcoal rot	Rabi- CSH 8(R)(T), M 35-1 (T). SPV104(T), CSV5R(T)
4. Pearl millet/ Bajra	Downy mildew: Ergot	MBH 110(R), Pusa 46(R), Pusa moti (T), RHR 1 (R), ICMS 7703 (R), WCC 75 (R), MBH110(T), WCC75(T)
5. Maize	Stalk rot, Pythium stalk rot, Ear rot, stalk rot and Foliar diseases	Rudrapur local, GS 2 Gs2 Deccan 101 Deccan 103 Deccan 103, GS2 Amber, Sona, Kisan

Breeding for Insect Resistance

All the plant species are attacked by insects, but the degree of damage to as well as the number of insect species attacking different crop species vary considerably. *For example,* cotton is attacked by more than 160 species of insect; of these about a dozen are major insect pests. Therefore, an effective pest control measure is the basic requirement for a good cotton crop; this is clear from the fact that about 60 per cent of all the pesticides used in the world are applied to cotton.

Once it was hoped that chemical control measure will effectively control, or even eliminate the insect pests. But the experience with pesticides has shown that such a hope was entirely misplaced. Extensive pesticide application (1) increases the cost of production of crops, (2) reduces the population of natural enemies (predators and parasites) of insect pests, (3) leads to the development of pesticide-resistant race of insects, and (4) pollutes the environment. From a practical view point, an insect resistant variety produces relatively larger yields of good quality than do susceptible varieties with the same level of initial infestation (insect attack) under comparable environmental conditions. Thus, insect resistance is a relative property and can be defined only in comparison to other more susceptible varieties; is can be defined as *'those heritable characteristics possessed by the plant, which influence, the ultimate degree of damage done by insects"* (Maxwell *et al.*, 1972).

Historical

The first report of insect resistance dates back to 1792, when Havens reported that wheat variety 'Underhill' was resistant to Hessian fly (*Mayetiola destructor*). In 1831, Lindley reported that apple variety 'winter majestic was resistance to wooly apple aphid (*Eriosoma langigerum*). In 1861, grape *Phylloxera vartifoliae* was controlled in France by the use of resistance root stocks introduced from U.S.A. This seems to be the first planned utilization of insect resistance. Subsequent developments have yielded a wealth of information on the various aspects of insect resistance and its commercial exploitation. As a result, breeding of insect resistant varieties is now on important objective in many national and international crop improvement programmers. In cultivation on pest resistant varieties is now considered a key component in the modern pest management practices.

Losses due to Insects

The loss in quality of the produce is more frequent and often of a greater concern than of a greater concern than that of quantity per se. Insect infestation leads to or more of the following directs damages: (1) reduced plant growth or stunting, (2) damage to leaf, stem, branch, flower dubs, flower, vegetative buds, fruits and seeds, (3) premature defoliation of leaves, and even (4) wilting of plants. The degree of damage depends primarily on the intensity of insect attack and the susceptibility of the host strain. In addition, insects also cause indirect damage. Many insects, *e.g.*, aphids, mites, white fly, *etc.*, transmit plant viruses, *i.e.*, serve as vectors of pathogens. Further, injuries caused by insect make the plant more vulnerable to attacks by fungal and bacterial pathogens.

The estimated global average loss due to insect pests in the potential yields all the crops is about 14 per cent (Cramer, 1967). The estimated losses in individual crops vary from 5 per cent in wheat to 26.7 per cent in rice, and still more in crops like cotton and sugarcane, in spite of the widespread use of insecticides and other control measures. The global average loss in most of the major crops exceeds 5 per cent. In India, losses due to insect pests range from 10 to 20 per cent. Losses due to insect pests are much more in tropical and subtropical than in temperate regions since high temperature and humidity promote insect activity and reproduction, in cases of severe insects attacks, the yield losses may be up to 90 per cent.

Mechanism of Insect Resistance

1. **Non preference/Non-acceptance/Antixenosis:** Colour, light penetration, hairiness, leaf angle, odour and taste.
2. **Antibiosis:** Adverse effect of host plant because benzyl alcohol, silica, DIMBOA, Aspartic acid, tanning, saponin, *etc.*
3. **Tolerance:** Ability to produce more than susceptible variety because rejuvenation potential, healthy leaf growth, flowering compensation potential and superiors plant vigour.

4. **Avoidance/Escape:** Because earliness or its cultivation in season where in selected population very low.

Basis of Insect Resistance

1. Morphological Factors

- ☆ Hairiness
- ☆ Colour of plant
- ☆ Solid stem
- ☆ Toughness tissues
- ☆ Others: Narrow lobes, leathery leaves, long pedicel, nectariless, frego bract *etc.*

2. Physiological Factors

- ☆ Osmotic concentration of cell sap.
- ☆ Leaf exudates (gummy/antibiotic), *etc.*

3. Biochemical Factor

Presence of Silica, benzyl alcohol, tannins, gossypol, spooning, DIMBOA, ethanol *etc.*

Genetics of Insect Resistance

1. Oligogenics

- ☆ Wheat- Green budge, Hessian fly resistance, *etc.*
- ☆ Rice- Plant hopper resistance
- ☆ Cotton: Jassids resistance
- ☆ Apple: Woolly aphids resistance, *etc.*

2. Polygenic

- ☆ Wheat- Leaf beetle resistance
- ☆ Rice-Stem bore resistance
- ☆ Maize- Ear worm, resistance
- ☆ Brassica – Aphid resistance, *etc.*

3. Cytoplasmic

- ☆ Maize- European corn borer resistance
- ☆ Lettuce- Root aphid resistance, *etc.*

Source of Insect Resistance

1. Cultivated varieties: *e.g.* In cotton, B 1007-Jassid resistance
2. Germplasm collection

3. Wild species
4. Induced mutation
5. Micro-organisms

Durable Resistance

Durability depends on;

1. Formation of new races/biotypes.
2. Genetic of resistance.
3. Morphological features of host plant.
4. Biochemical substances associated with resistance.

Breeding Methods

1. Plant Introduction
2. Selection
3. Hybridization
4. Mutation
5. Transgenic breeding

Merits

- Controlling loses.
- Reduction in cost of production and increase the cost benefit ratio.
- Non toxic to man.
- Only selection for some diseases – rusts, wilts, smuts, nematode and bacterial blights.

Demerits

- Long term (10-15 year).
- Resistance to one pest leads to the susceptibility to another pest.
- Inter-specific gene transfer posses many problems.
- Expensive method.
- Resistance variety has lower yield and poor quality

Achievements

- *Phylloxera* resistance variety of grapes.
- 23 Hessian fly resistance wheat varieties
- Stem sawfly resistant wheat variety- Rescue (1946)
- Green bug resistance barley variety– Will
- Alfalfa aphid resistance- Cody, Moapa and Zia, *etc.*

22

Plant Breeders/Farmers Rights, Farmers Privilege and EDVs

Plant breeder's rights are the rights granted by the government to a plant breeder, originator or owner of a variety to exclude others producing or commercializing the propagating material of that variety for a minimum period of 15-20 years. A person holding PBR title to a variety can authorize other interested persons/organization to produce and sell the propagating material of that variety. He should set reasonable terms for such transfers of PBR titles or for the sale of the propagating materials; otherwise the government can grant licenses of the titles in public interest. It is important that the object of protection in PBR is the variety, and that genetic components and the breeding procedures are not protectable. In addition, PBR systems also contain some form of 'breeders' exemption and 'farmers' privilege.

Historical

A plant act to provide patents on plants was first introduced in Germany in 1866. In the beginning of the present century, legislations for patenting plants were subsequently introduced in other countries, including U.S.A. where provisions for patenting varieties of asexually propagated crops were made. The International Organization for Plant Variety Protection (ASSINSEL) was established in 1938 with the objective of persuading government of different countries for introducing laws to protect plant varieties. Several countries, particularly in Europe, developed their own system of PBR.

The most significant event in the development of PBR systems was the effort to harmonize PBR laws of different countries through UPOV (Union Internationale pour la Protection des Obstentions Vegetales, *i.e.* International Union for Protection of New Plant Varieties). The first UPOV convention was signed in 1961 in Paris.

Requirements for PBR

Under the provision of UPOV 1991 act, a plant variety must satisfy the following four criteria for protection: (1) Novelty, (2) Distinctiveness, (3) Uniformity and (4) Stability.

The Extent of Protection by PBR

The previous of UPOV 1991 act offers the following protections to the concerned variety.

(i) Production for commercial purposes, offering for sale and selling all material becomes the exclusive right of the holder of the PBR-title.

(ii) A grower may be allowed to reserve a portion of his harvest for use as seed for his own next crop without the permission of the holder of the PBR-title. This is called farmers' exemption; but a farmer who does this is not allowed to sell such material to anyone else.

(iii) Exchange of propagating material of different cultivars between farmers is not allowed.

(iv) The minimum period of protection prescribed is 20 years. Several UPOV member states have established a longer period of protection of 20-25 years, with 30 years in France for inbred lines of maize, and for clovers and a few grasses.

(v) The use of propagating material from a protected cultivar for scientific purposes is not dependent on permission of the holder of the PBR-title.

(vi) The use of protected cultivar for the creation of genetic variability for plant breeding purposes is not dependent on permission of the holder of the title.

(vii) PBR protection covers the new variety, but does not protect the parents of the variety, except in the case of hybrid varieties.

Breeder's Exemption

Under PBR regime, the use of material of a protected variety (the initial variety) for the development of the breeding who developed them, and the holder of PBR-title of the initial varieties will have no claim to it. This provision is called breeders' exemption. Under the UPOV 1978 Act, all new varieties evolved using protected varieties were exempted from protection under this provision. But UPOV 1991 Act, has somewhat limited the scope of breeder's exemption by bringing *'essentially-derived varieties'* under the cover of PBR protection granted to the initial variety. An essentially derived variety (EDV) has been defined as a variety predominantly derived from the initial variety, which retains the expression of the essential characteristics from the genotype or combination of genotype of the initial variety. Thus a variety produced by, say mutation or transfer through backcross method/ integration by genetic transformation of a single gene will be considered as an 'essentially-derived variety', and will be protected under the PBR-title granted to the initial variety. The breeder of such a variety will, therefore, be required to obtain permission from the PBR-title holder of the initial variety.

Table 47: A Comparison of UPOV Act (1978) with UPOV Act (1991), PPVFR Act (2001) and Patents

Feature	*UPOV Act (1978)*	*UPOV Act (1991)*	*PPV & FR Act (2001)*	*Patent*
Protection coverage	Plant varieties of nationally defined plant species	Plant varieties of all plant genera and species	Varieties of nationally specified genera and species	Inventions
Requirements for protection	1. Distenctness 2. Uniformity 3. Stability	1. Novely 2. Distinctiveness 3. Uniformity 4. Stability	1.Novely 2. Distinctiveness 3. Uniformity 4. Stability	1. Novelty 2. Inventiveness 3. Nonobviousness 4. Industrial application and usefulness
Duration of protection	Minimum 15 yr	Minimum 20 yr	Maximum 15 yr for extant varieties and new varieties of crops; 18 yr for varieties of trees and vines	17-20 yr (OECD)+
Scope of protection	Commercial use of the reproductive material of protected variety	Commercial use of all material of the protected variety	Commercial use of all material of the protected variety	Commercial use of protected subject matter
Breeders' protection	Yes	Yes, except for essentially derived varieties	Yes, except for essentially-derived varieties, and use as parents of hybrid varieties	No
Farmers' privilege	Yes, (in practice)	Optional; left to the national laws	Yes, as 'farmers' rights'; more extensive than in UPOV Act (1978)	No

*PPVFR Act, 2001,The protection of plant varieties and Farmers' Right Act, 2001.

+ OESD, Organization for Economic Cooperation and Development.

Farmer's Privilege

PBR systems generally allow the farmers to use the material of a protected variety produced on their farm for planting of their new crop without any obligation to the PBR title holder. This exemption is usually referred to as farmers privilege. Under the UPOV 1978 Act, there was explicit provision for farmer's privilege. But in the proposed UPOV 1991 Act, this privilege has been made 'optional' and each UPOV member state can either allow or disallow this privilege. It should be clearly understood that farmer's privilege applies to the use of seed produced by a farmer for sowing 'his own' fields. PBR, however, does not allow farmers to exchange seeds of protected varieties produced on their farms.

Farmer's privilege is a very important provision for countries like India, where over 90 per cent of the total cropped area is sown by seeds produced by the farmers themselves. In addition, a majority of the farmers are poor and will be subject to unjust economic burden if they are forced to pay a royalty on the seed produced and used by them. The infrastructure of seed industry does not allow the option of use of new seed every year even if the farmers could be made, by some miracle, capable of the same.

Farmer's Rights

Agriculture began some 10,000 years ago. During this vast period of time genetic resources have been selected, developed, used and conserved by farmer families and farming communities of, particularly, the gene-rich developing countries. These same materials have been and are being collected, conserved and used as raw materials to evolve the modern high yielding varieties of various crops. Seed of these improved varieties earn huge profits for the seed corporation.

It has been argued that farmers should be allowed a share in this profit in recognition of their contribution by way of the development of germplasms of the various crops. This has been recognized by FAO (Resolution No. 5/89) as farmer's rights, which arise from the past, present and future contribution of farmers in conserving, improving and making available plant genetic resources (PGR), particularly in the centers of origin/diversity. It has been emphasized that the farmer's rights should be obligatory and should not be relegated as privileges.

Benefits from PBR

1. The opportunity to breeders of obtaining profits from varieties developed by them will act as an incentive in promoting plant breeding.
2. It encourages private companies to invest in plant breeding activities.
3. It will enable access to varieties developed in other countries and protected by IPR laws.
4. Increased competition among various organization engaged in plant breeding is likely to be beneficial to both the farmers and the nation.

Disadvantages from PBR

1. PBR will encourage monopolies in genetic material for specific traits.
2. It suppresses free exchange of genetic material and may encourage unhealthy practices.
3. The holder of PBR-title may produce less than the demand in order to increase prices for achieving more profit.
4. Farmer's privilege to re-sow the seed produced by him may become gradually diluted/eliminated.
5. PBR may result in increased cost of seed, which will be burdensome to the poor farmers of India, and would limit the benefits from new varieties to a small segment of rich farmers.

23

Participatory Plant Breeding (PPB)

Participatory plant Breeding refers to breeding program based on set methods that involve close farmer-researcher collaboration to bring out crop improvement and it is logical extension of participatory variety selection. Farmer's participatory varietal selection (FPVS) has emerged as the best method to identify farmers' preferred crop varieties and their popularization. FPVS is one of the plants breeding approach in recent days which gaining importance among the scientist/farming community for rapid spread and adoptability of improved cultivars to the farming community. As it is a participatory approach, this method includes partners like farmers, scientist/researchers, seed agencies, extension workers, traders, millers and agriculture based industries and allied departments. Among them farmers are fore most important partners. The variety identified by this method needs to satisfy the needs of all the partners; hence it ensures the increase in productivity and profitability of the crop.

FPVS Implementation Approach

This method will be implemented with the new approach called *MOTHER-BABY* trails. This approach serves as an evaluation tool for farmer's participatory varietal selection method. Mother-Baby trails are planned by scientists and managed by farmers. Among the beneficiary farmers, one farmer will be selected to carry out mother trail, which means all the varieties available in the method are evaluated in single plot. Rest of the farmers will conduct baby trails in paired comparison passion, which means each farmer evaluated two genotypes in combination along with their local checks (Fig. 38 and 39). During the experimentation period scientist/

researcher will visit experimental plots and provided necessary information to farmers timely. All the beneficiary farmers will be provided with "experimental kit" which includes critical inputs like seeds, plant protection chemicals and more importantly micronutrients.

Types of Participatory Plant Breeding

1. Formal led participatory plant breeding
2. Farmer led participatory plant breeding

Different Models of PPB

Based on Stages of Involvement

1. Traditional farmer breeding
2. Complete participatory breeding
3. Efficient participatory breeding
4. Participatory varietal selection (PVS)
5. Scientist plant breeding

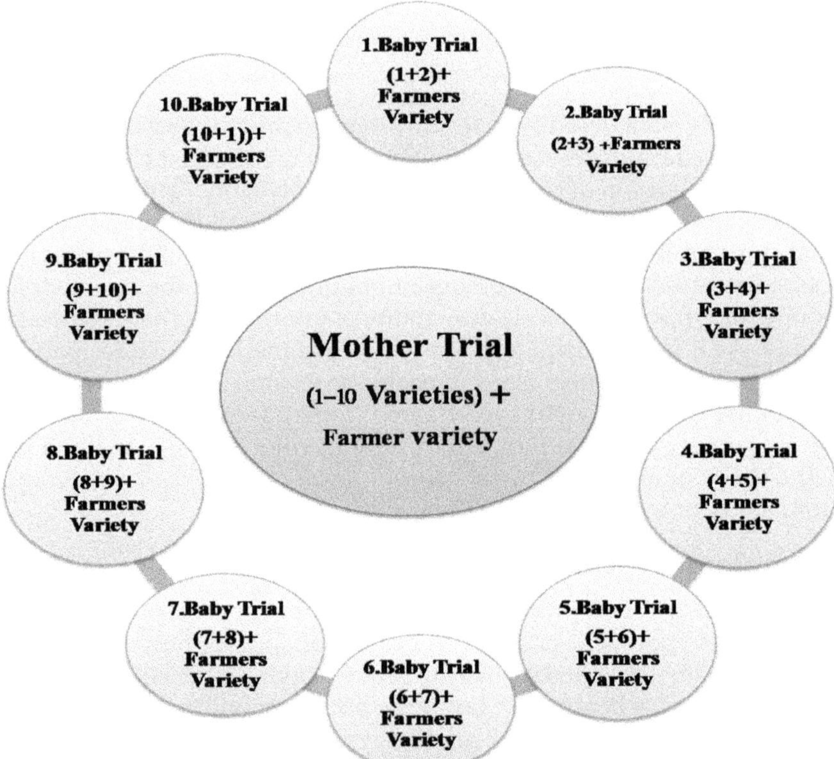

Figure 42: Schematic Representation of Mother-Baby Trial Approach with 10 Test Genotypes.

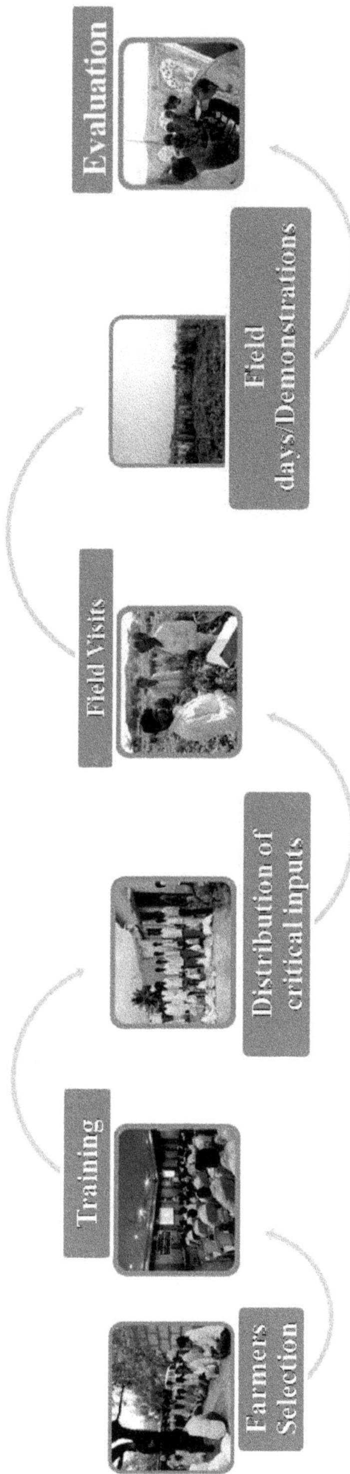

Figure 43: Schematic Representation of FPVS Breeding Approach.

Objectives

- ✰ Enhancing biodiversity and germplasm conservation.
- ✰ Enhanced adoption by farmer.
- ✰ Benefit to specific end user.
- ✰ Farmer skill development – selection and seed production

Table 48: Different Models of PPB: Based on Stage of Involvement

Steps	*Participation*									
	Traditional Famer Breeding		*Complete Participatory Breeding*		*Efficient Participatory Breeding*		*Participatory Varietal Selection (PVC)*		*Scientist Plant Breeding*	
	F	*S*	*F*	*S*	*F*	*S*	*F*	*S*	*F*	*S*
Selection of source germplasm	✓		✓	✓	✓	✓		✓		✓
Trait development (Pre-breeding)	✓		✓	✓		✓		✓		✓
Selection of source germplasm	✓		✓	✓		✓		✓		✓
Varietal evaluation	✓		✓	✓	✓	✓	✓	✓		✓

PVC + PPB: Participatory Crop Improvement; F: Farmer; S: Scientist.

Advantages

- ✰ Improved local adaptation.
- ✰ Promotion of genetic diversity.
- ✰ Increased breeding efficiency.
- ✰ Empowerment of rule communities (self-reliance).
- ✰ Gender empowerment.

Shortcoming

- ✰ High overall cost for breeding programme.
- ✰ High cost for participating farmers.
- ✰ Additional training needed to scientist.

Challenges

1. Technical: *e.g.* seed certification
2. Economic: *e.g.* high cost, more resources needed
3. Institutional: *e.g.* Ownership issue, benefit sharing issue

Participatory Plant Breeding should be used in following conditions

1. Tribal and remote area.
2. Highly variable environment.
3. Risk prone, complex and low input-agriculture.
4. End users are diverse and locally unique.
5. Crop of local interest *i.e.* minor or underutilized crop.
6. End user require uncommon trait.
7. Unusual combination of common traits.
8. Self–pollinated crop.

Successful Examples

- ☆ Kalinga-3 (Rice), GDM-6 (Maize), Mother body trial (Maize).
- ☆ Mariana Baninger- African Maize Stress Project, CIMMYT got King Baoudin Award from CGIAR for this innovative project.

24

Modern Plant Breeding Strategies

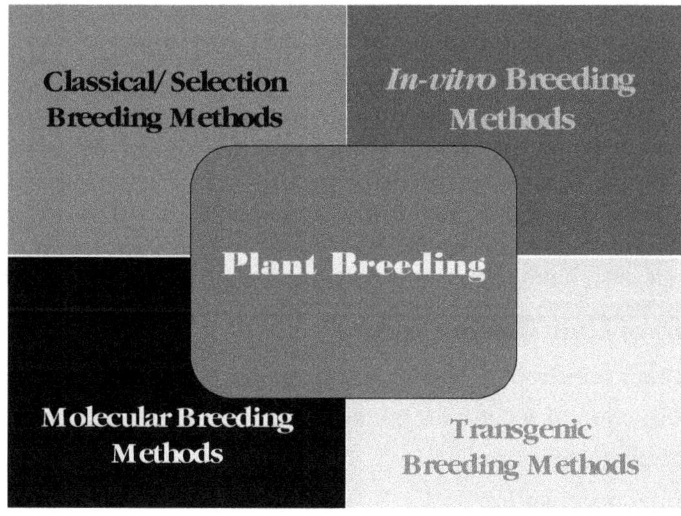

Figure 44: Modern Plant Breeding Strategies.

A. Conventional/Classical/Traditional Breeding Methods

Important conventional breeding methods are listed below.

- ☆ Selection
- ☆ Hybridization

- ☆ Mass Selection
- ☆ Bulk Breeding Method
- ☆ Single-Seed Descent
- ☆ Pedigree Breeding
- ☆ Recurrent Phenotypic Selection
- ☆ Synthetic and Composite Cultivars
- ☆ Developing Hybrid Cultivars
- ☆ Development of Clonal Cultivars
- ☆ Mutagenesis
- ☆ Polyploidy
- ☆ Heterosis *etc.*

NEW GENETIC APPROACHES

B. Tissue Culture (*in-vitro*) Breeding Techniques

The term "Plant tissue culture" broadly refers to the *in vitro* cultivation of plant parts under aseptic conditions. Such parts as meristems, apices, axillary buds. Young inflorescence, leaves, stems, and roots have been cultured. A controlled aseptic environment and suitable nutrient medium are the two chief requirements for successful tissue culture. These essential nutrients include inorganic salts, a carbon and energy source, vitamins and growth regulators.

The basic technology can be divided into five classes, depending on the material being used: Callus, organ, meristem, and protoplast and cell culture. The technique of embryo, ovule, ovary, anther and microspore culture are used and can yield genotypes that cannot easily be produced by conventional methodology. A variety of techniques (embryo culture, apical culture, anther or pollen culture, haploid production, micro-propagation, somaclonal variation, protoplasts, somatic embryogenesis, *etc.*) have been developed under the title of tissue culture.

Application of Cell/Tissue Culture

1. Mutant selection
2. Production of secondary metabolites or biochemical production.
3. Biotransformation
4. Clonal propagation
5. Somaclonal variations

Embryo Culture

For embryo culture, embryos are excised from immature seeds, usually under a 'hood', which provides a clean aseptic and sterile area. Sometimes, the immature seeds are surface sterilized and soaked in water for few hours, before the embryos are excised. The excised embryos are directly transferred to a culture dish or culture tube containing synthetic nutrient medium.

Application of Embryo Culture

1. Recovery of distant hybrids.
2. Recovery of haploid plants from Interspecific crosses.
3. Propagation of orchids.
4. Shortening the breeding cycle
5. Overcoming dormancy.

In addition ovule and ovary can also be cultured.

Meristem Culture

In attempts to recovery pathogen free plants through tissue culture techniques, horticulturists and pathologists have designated the explants used for initiating cultures as 'shoot –tip', tip, meristem and meristem tip. Therefore, the term 'meristem' or meristem-tip' culture is preferred for in vitro culture of small shoot tips. The in vitro techniques used for culturing meristem tips are essentially the same as those used for aseptic culture of plant tissues. Meristem tips can be isolated from apices of the stems, tuber sprouts, leaf axils, sprouted bunds o cuttings or germinated seeds.

Application of Meristem Culture

1. Vegetative propagation
2. Recovery of virus free stock.
3. Germplasm exchange
4. Germplasm conservation

Anther or Pollen Culture

Angiosperms are diploid the only haploid stage in their life cycle being represented by pollen grains. From immature pollen grains we can sometimes raise cultures that are haploid. These haploid plants have single completes set of chromosomes. The usual approach in anther culture is that anthers of appropriate development stage are excised and cultured so that embryogenesis occurs. Alternatively pollen grains may be removed form the anther, and the isolated pollen is then cultured in liquid medium. Cultured anthers may take upto two months to develop into plantlets.

Application

Pollen culture or anther culture is useful for production of haploid plants. Similarly, haploid plants are useful in plant breeding in variety of ways as follows:

1. Releasing new varieties through F1 double haploid system.
2. Selection of mutants resistant to diseases.

3. Developing asexual lines of trees or perennial species.
4. Transfer of desired alien gene.
5. Establishment of haploid and diploid cell lines of pollen plant.

Haploid and Double Breeding

Establishing true breeding, homozygous, lines is an essential part of developing new cultivars in many crop species. These homozygous lines are used either as cultivars in their own right (*i.e.* for inbreeding crop species) or as parents in hybrid variety development. Traditionally, plant breeders have used the process of selfing or mating between close relatives to achieve homozygosity, a process that is time-consuming. Therefore the opportunity to produce plants from gametic, haploid cells has been the goal of many plant breeders as this technique would produce instant inbred lines once the chromosomes of the haploids are doubled.

The genetic phenomenon critical to obtaining homozygous lines is the formation of haploid gametes by meiosis. During this type of cell division, the chromosome number is halved and each chromosome is represented only once in each cell (assuming the species is basically a diploid one). If such gametic, haploid cells can be induced to develop into plantlets a haploid plant can develop which can then be treated to encourage its chromosomes to double, to produce a completely homozygous line (a doubled haploid). There are a number of methods of haploid induction that are not directly related to tissue culture but the most widely applicable are via the culture of anther or microspore (immature pollen grains) *in vitro*.

Micro-propagation

"Clonal propagation through tissue culture technique is called micro propagation." In other words regeneration of whole plant through tissue culture is popularly called micro-propagation. Micro Propagation can be achieved in a short time and space. Thus, it is possible to produce plants in large numbers starting from a single individual. Use of tissue culture for micro-propagation was initiated by G. Morel (1960) in orchid.

In vitro multiplication of breeding lines can have two main benefits (particularly in clonal species) in relation to plant breeding programmes.

- ✰ Plants propagated *in vitro* can generally be initiated and maintained in a disease-free state, and so can be used to help maintain stocks of breeding lines; facilitate long-term germplasm storage; and facilitate international exchange of material.
- ✰ Short generation times and fast growth means that rapid increases in plant number can be readily achieved.

Both the above have particular importance to clonal crops, because these tend to have a relatively low multiplication rate as a result of their vegetative mode of propagation and are particularly susceptible to viral and bacterial diseases, which tend to be multiplied and transmitted through each clonal generation.

Somaclonal Variation

According to Larkin and Scowcroft (1981), "Somaclonal variation is the genetic variability which is regenerated during tissue culture" or plant variants derived from any form of cell or tissue cultures.

It has been recognised that all plants regenerating from the tissue culture are not exactly the replicas of a parental form. Phenotypic variation is frequently observed amongst regenerated plants. Phenotypic changes are associated with genetic changes of an organism. The frequently genetic changes could result from mutations, epigenetic changes or combination of both mechanisms. Since, genetic mutations are irreversible and likely to persist in the progeny of regenerated plants whereas epigenetic changes are not transmitted by sexual reproduction. It has been found that cells or tissues in cultures undergo frequent genetic changes resulting in new phenotypes on organogenesis or embryogenesis. Plant cell and tissue culture provides increased genetic variability. Variants selected in tissue cultures have been referred to as "calliclones" or "Protoclones".

Gametoclonal Variations

Evans *et.al.* prefer the term "Gametoclonal variation" for variant clones specifically raised from gametic or Gametophytic cells.

C. Plant Biotechnology

The origin of Biotechnology can be traced back to prehistoric times, when microorganisms were already used for processes like fermentation. In 1920's *Clostridium acetobutylicum* was used by Chaim Weizman for converting starch into butanol and acetone, latter was an essential component of explosive during World War- II. This raised hopes for commercial production of useful chemicals through biological processes, and may be considered as the first rediscovery of biotechnology in the present century. Similarly, during World War-II (in 1940's), the production of penicillin (as an antibiotic discovered by Alexaner Flemming in 1929) on a large scale from cultures of *Penicillium notatum* marked the second rediscovery of biotechnology. The third rediscovery of biotechnology is its recent reincarnation in the form of recombinant –DNA technology, which led to the development of a variety of gene technologies and is thus considered to be greatest scientific revolution of this century.

Definitions

1. Biotechnology is the application of biological organisms, system or processes to manufacturing and service industries.
2. Biotechnology is the integrated use of biochemistry, microbiology and engineering science in order to achieve technological application of the capabilities of micro-organism, cultured tissue cells and part thereof.
3. Biotechnology is "a technology using biological phenomenon for copying and manufacturing various kinds of useful substances."

4. Biotechnology is "the controlled use of biological agents such as micro-organisms or cellular components for beneficial use. (U.S National Science Foundation)

Plant Biotechnology is about

- ☆ Manipulating plants for the benefit of mankind
- ☆ A precise process to improve plants.
- ☆ More Food
- ☆ Better Food
- ☆ Better environment

Why Plant Breeding needs Plant Biotechnology?

- ☆ Improvement in the yield
- ☆ Improvement in the insect and disease resistance
- ☆ Improvement in the quality
- ☆ Herbicide resistance
- ☆ Resistance to abiotic stresses
- ☆ Rapid and accurate technology
- ☆ No barrier for gene transfer
- ☆ Industrial products
- ☆ Application of genetics principles for improvement
- ☆ Directed manipulation of the genotype at the DNA sequence level, and introduction of new genes.

Transgenic Breeding

The stable introduction of specific genes into plants represents one of the most significant developments affecting the production of crop species in a continuum of advances in agricultural technology. The progress in this area has depended largely on the tissue culture systems having been developed which, at least, initially, provide an amenable vehicle for the transformation induction.

The term transformation comes from that used for a much longer period, bacterial transformation, in which DNA has been successfully transferred from one isolate to another or another species of bacteria, and integrated into the genome. It was shown that the stably transformed bacteria then expressed the new genes and displayed appropriately altered phenotypes. In eukaryotes, transformation has a further complicating dimension, at least in many plants' breeding contexts. The transforming DNA must not only be integrated into a chromosome, it must be a chromosome of a cell, or cells that will develop into the germline. Otherwise the transformation will not be passed on to the next generation.

Using plant transformation techniques it is possible to transfer single genes (*i.e.* simply inherited traits) into plants, to have such transgenes expressed and for them to function successfully. Theoretically at least, specific genes can be transformed from any source into developed cultivars or advanced breeding lines in a single step. Plant transformation, therefore, would appear to allow plant breeders to bypass barriers that limit sexual gene transfer and to exchange genes (and traits) from unrelated species between which sexual hybridization is not possible. These recombinant DNA techniques, apparently, allow breeders to transfer genes between completely unrelated organisms. For example, bacterial genes can be transferred and expressed in plants. This appears to break the barrier that sexual reproduction generally imposes. However, as we learn more about the DNA, and hence the genes involved, the perspective of the picture changes somewhat, with increasing direct evidence of the presence in different species of the same basic gene, or clear variants of it, and demonstrations of the greater conservation of genetic material (synteny) during evolution than we expected. Also, we are being reminded of the existence of parallel natural processes for much of what we regard as novel. For example, bacteria, viruses and phages already have successfully evolved mechanisms to transfer genes just in the way we regard as being so alien! But clearly, the new techniques are allowing modern plant breeders to create new variability beyond that existing in the currently available germplasm on a different scale and in a different time frame from that which was previously possible.

D. Molecular Breeding

Although plant breeders have practiced their art for many centuries, genetics is a subject that really only came of age in the twentieth century with the rediscovery of Mendel's work. Since then research in genetics has covered many aspects of the inheritance of qualitative and quantitative traits, but plant breeders usually still have little, or no, information about:

- ☆ The locations of many of these loci in the genome or on which chromosome they reside;
- ☆ The number of loci involved in any trait;
- ☆ The relative size of the contribution of individual alleles at each loci on the observed phenotype, except where there is an obvious major effect (*e.g.* height and dwarfing genes).

The Idea of using Markers

The idea of associating easily visualized markers in plants with loci affecting qualitative and quantitative variation in traits of interest to plant breeders is not new, and was first proposed by Sax in 1923. The basic idea is relatively simple. If a trait or characteristic is difficult to score (*e.g.* it shows continuous variation; assessment is detailed and time consuming; or the trait is only expressed after several years of growth), an easily scored marker that was determined by a locus closely associated with that affecting the character would be an attractive way to monitor the locus of interest.

The characteristics of a good marker system include the following:

1. The markers are easy, quick, and inexpensive to score.

2. The markers themselves have no deleterious effects on fitness and no effects on other traits, including undesirable epistatic interactions with any other traits.
3. There is a high level of variation exhibited.
4. They are stable in expression over environments.
5. Assessment can be made early in the development of the plant (seedling level), and/or in tissue culture.
6. The scoring should be nondestructive in terms of the whole plant, so that desirable individuals can be selected and still grown to maturity.
7. Co-dominance in expression of the alternative alleles, so that heterozygotes can be differentiated from either homozygous dominant genotype.

Types of Marker Systems

The types of markers that can and have been used in plant breeding include:

1. **Morphological Markers**: Which are basically those that you see by simply looking at a plant's phenotype, including characters such as pigmentation, dwarfism, leaf shape, absence of petals, *etc.*
2. **Biochemical Markers:** Such as isozyme markers. Isozymes (isoenzyme) are variant forms of an enzyme, which are functionally identical but can be distinguished by electrophoresis – *i.e.* when placed in an electric field. Under these circumstances the different forms of the enzyme will migrate to different points in the electric field depending on their charge, size and shape.
3. **Molecular Markers:** These represent the variation that is present and can be detected at the level of DNA. There are basically two systems (PCR and non-PCR based) by which molecular markers are generated and their distinction need not detain us, but it is worth pointing out that molecular markers are simply differences in the DNA between individuals, groups, species taxa *etc.* Clearly the type and level of variation in DNA that we would want to examine is different depending on what level of distinction we are interested in and what questions we are answering.

But the main characteristics of molecular markers are that: they are a ubiquitous form of variation; they are free from environmental influence; they show high levels of polymorphism; they have no discernible effects on the phenotype; and they can be detected using only small pieces of tissue.

Given the above characteristics of molecular markers, particularly their relatively unlimited numbers, it is no surprise that the advent of the possibilities of molecular markers in the 1990s was greeted with some excitement and is seen as providing a major change in the potential to exploit the ideas for using markers advocated some 70 years earlier.

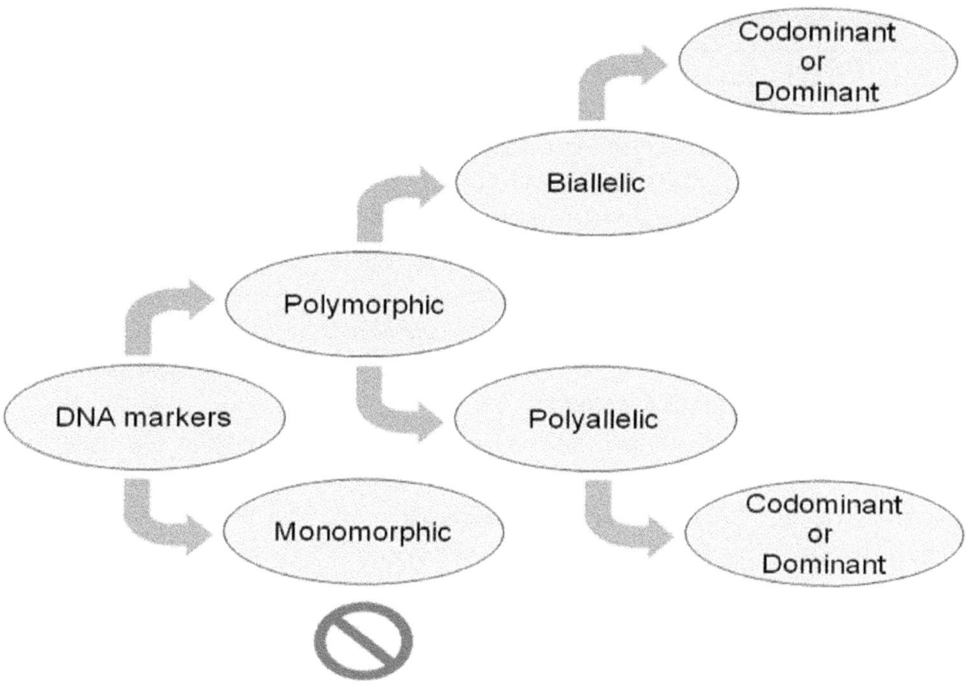

Figure 45: Classification of DNA Markers Based on Discreetness.

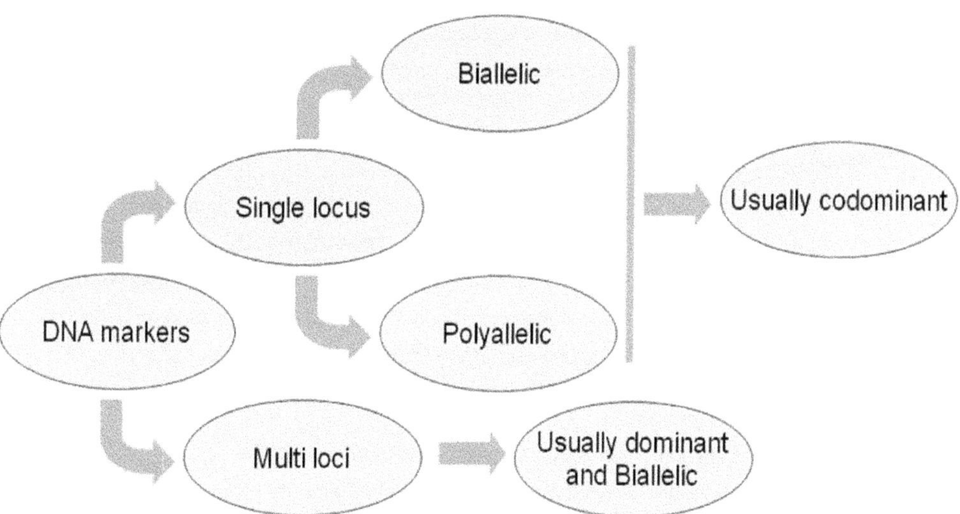

Figure 46: Classification of DNA Markers Based on Location on the Genome.

Table 49: Classification of DNA Markers Based on their Time of Discovery

Year	*Acronym*	*Nomenclature*	*Reference*
A. FIRST GENERATION DNA MARKERS			
1974	RFLP	Restrition Fragment Length Polymorphism	Grodzicker *et al.* (1974)
1985	VNTR	Variable Number Tandem Repeats	Jeffreys *et al.* (1985)
1986	ASO	Allele specific oligonucleotides	Saiki *et al.* (1986)
1988	AS-PCR	Allele specific polymerase chain reaction	Landegren *et al.* (1988)
1988	OP	Oligonucleotide polymorphism	Beckmann (1988)
1989	SSCP	Single Stranded Conformational Polymorphism	Orita *et al.* (1989)
1989	STS	Sequence Tagged Site	Olsen *et al.* (1989)
B. SECOND GENERATION DNA MARKERS			
1990	RAPD	Randomly Amplified Polymorphic DNA	Williams *et al.* (1990)
1990	AP-PCR	Arbitrarily Primed Polymerase Chain Reaction	Welsh and McClelland (1990)
1990	STMS	Sequence Tagged Microsatellite Sites	Beckmann and Soller (1990)
1991	RLGS	Restriction Landmark Genome Scanning	Hatada *et al.* (1991)
1992	CAPS	Cleaved Amplified Polymorphic Sequence	Akopyanz *et al.* (1992)
1992	DOP-PCR	Degenerate OligonucleotidePrimer - PCR	Telenius (1992)
1992	SSR	Simple Sequence Repeats	Akkaya *et al.* (1992)
1993	MAAP	Multiple Arbitrary Amplicon Profiling	Caetano-Anollés *et al.* (1993)
1993	SCAR	Sequence Characterized Amplified Region	Paran and Michelmore (1993)
C. NEW GENERATION DNA MARKERS			
1994	ISSR	Inter Simple Sequence Repeats	Zietkiewicz *et al.* (1994)
1994	SAMPL	Selective Amplification of Microsatellite Polymorphic Loci	Morgante and Vogel (1994)
1994	SNP	Single Nucleotide Polymorphisms	Jordan and Humphries (1994)
1995	AFLP (SRFA)	Amplified Fragment Length Polymorphism (Selective Restriction Fragment Amplification)	Vos *et al.* (1995)
1995	ASAP	Allele Specific Associated Primers	Gu *et al.* (1995)
1996	CFLP	Cleavase Fragment Length Polymorphism	Brow (1996)
1996	ISTR	Inverse Sequence-Tagged Repeats	Rhode (1996)
1997	DAMD-PCR	Directed Amplification of Minisatellite DNA-PCR	Bebeli *et al.* 1997
1997	S-SAP	Sequence-Specific Amplified Polymorphism	Waugh *et al.* (1997)
1998	RBIP	Retrotransposon Based Insertional Polymorphism	*Flavell et al.* (1998)
1999	IRAP	Inter-Retrotransposon Amplified Polymorphism	Kalendar *et al.* (1999)
1999	REMAP	Retrotransposon-Microsatellite Amplified Polymorphism	Kalendar *et al.* (1999)
1999	MSAP	Methylation Sensitive Amplification Polymorphism	
2000	MITE	Miniature Inverted-repeat Transposable Element	Casa *et al.* (2000)
2000	TE-AFLP	Three Endonuclease AFLP	van der Wurff *et al.* (2000)
2001	IMP	Inter-MITE polymorphisms	Chang *et al.* (2001)
2001	SRAP	Sequence-related amplified polymorphism	Li and Quiros (2001)

Table 50: Comparison of DNA Markers for the Efficiency

Markers	*RFLP*	*Mini Satellites*	*RAPD*	*SSR*	*ISSR*	*SSCP*	*CAPS*	*SCAR*	*AFLP*
Abundance	H	M	H	H	H	L	L	L	H
Level of polymorphism	M	M	M	H	M	L	M	M	M
Locus specificity	Yes	No	No	Yes	No	Yes	Yes	Yes	No
Co-dominance of alleles	Yes	No	No	Yes	No	Yes	Yes	No	No
Reproducibility	H	H	L	H	M	M	H	H	H
Labour-intensity	H	H	L	L	L	M	M	L	M
Technical demands	H	H	L	M	M	M	M	L	M
Operational costs	H	H	L	M	M	M	M	L	M
Development costs	M	M	L	H	L	H	H	H	L
Quantity of DNA required	H	H	L	L	L	L	L	L	M
Amenability to automation	No	No	Yes	Yes	Yes	No	Yes	Yes	Yes

H: High; M: Medium; L: Low.

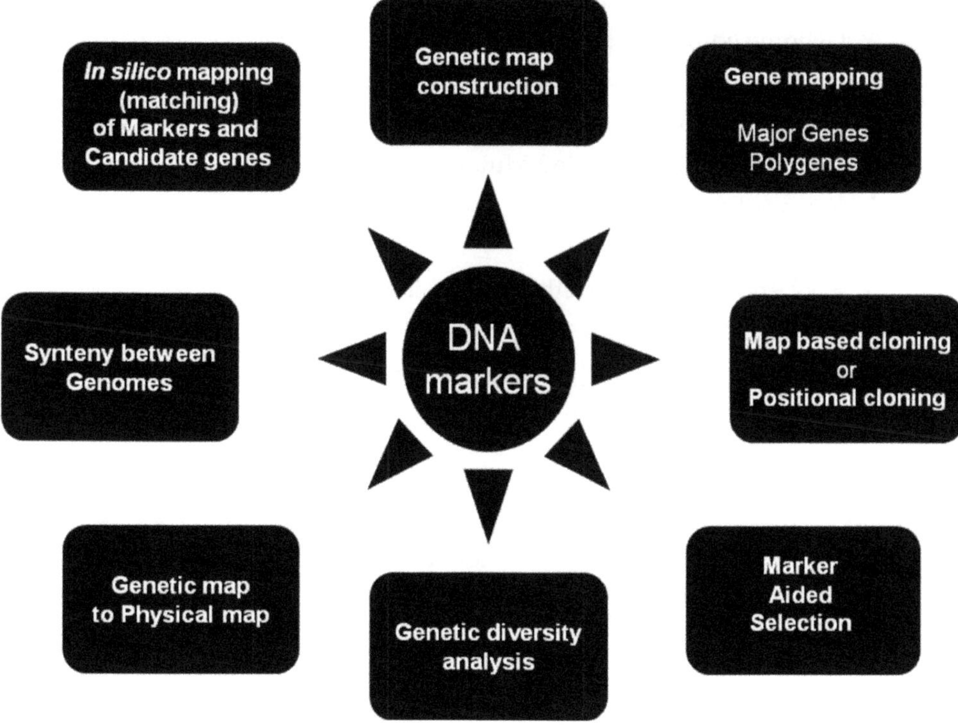

Figure 47: Applications of Molecular Markers.

Table 51: Comparison of Genetic Modification Techniques

Particulars	*Selective Breeding*	*Mutation Breeding*	*Transgenic Breeding*
Level	Whole genome	Molecule	Molecule
Precision	Thousands of genes	Unknown	Single gene
Certainty	Genetic change poorly characterized	Genetic change poorly characterized	Gene function well understood
Limits	Between species and genera	Not applicable	No limitations
Linkage drag	Present	Absent	Absent
Improvement of traits	Possible (qualitative), with great difficulty (quantitative)	Hit of miss method	Possible
Safety issues	Not required	Not required	Required

Other Important Novel Plant Breeding Approaches

- ☆ Agro-infiltration
- ☆ Cisgenic Modifications
- ☆ Genome editing techniques
- ☆ Grafting non-GM Scion onto GM Rootstock
- ☆ Intragenic Modifications
- ☆ Oligonucleotide- Directed Mutagenesis
- ☆ Reverse Breeding
- ☆ RNAi
- ☆ RNA-induced DNA Methylation Gene Silencing
- ☆ Targeted Mutagenesis
- ☆ Virus Induced Gene Silencing
- ☆ Xenogenic Modifications, *etc.*

Future Potential

Plant breeding will continue to be highly dependent on classical techniques but will undoubtedly increase in efficiency and effectiveness by the addition of these new approaches, which will be used in parallel with the more classical ones. Thus the future will see the range of techniques expanding in such a way as to maximize their benefits by their integrated exploitation.

References

Agrawal, P.K. and Dadlani, M. (eds.) 1987. Techniques in Seed Science and Technology. South Asian Publishers, New Delhi.

Allard, R.W. 1960. Principles of Plant Breeding. John Wiley and Sons, New York.

Allard, R.W. and Bradshaw, A.D. 1964. Implications of Genotype-Environmental Interactions in Applied Plant Breeding. Crop Science. 4: 503-508.

Banga, S. S. and Banga, S. K. (eds) 1998. Hybrid Cultivar Development. Narosa Publication House, New Delhi.

Blum, A. 1988. Plant Breeding for Stress Environments. CRC Press, Inc., Boca Raton, Florida.

Borlaug, N. E. 1959. The use of Multibreed or Composite Varieties to Control Airbone Epidemic Disease of Self-pollinated Crop Plants. *Proceedings of First International Wheat Genetics Symposium*, pp. 12-26.

Brim, C. A. and Stuber, C. W. 1973. Application of Genetic Male Sterility to Recurrent Selection Schemes in Soybeans. *Crop Sci.* 13: 528-530.

Brown, A.H.D. and Allard, R.W. 1971. Effect of Reciprocal Recurrent Selection for Yield on Isozyme Polymorphism in Maize. *Crop Sciences*. 11: 888-893.

Chopra, V. L. 1990. In-vitro Genetic Modification of Crop Brassicas: A Case Study of Application of Biotechnology for Crop Improvement. In (J. P. VERMA and A. VERMA Eds.) Technology Blending and Agrarian Prosperity. Malhotra Publishing House, New Delhi.

Chopra, V. L. and Paroda, R. S. (eds.) 1986. Approaches for Incorporating Drought and Salinity Resistance in Crop Plants. Oxford and IBH publ. Co. Pvt. Ltd., New Delhi.

Cockerham, C. C. and Matzinger, D. F. 1985. Selection responses based on selfed progenies. *Crop Sciences*. 25: 483-488.

Comstock, R. E. Robinson, H. F. and Harvey, P. H. 1949. A Breeding Procedure Designed to Make Maximum use of both General and Specific Combining Ability. *Agronomy Journal*. 41: 360-367.

Donald, C. M. and Hamblin, J. 1983. The Convergent Evolution of Annual Seed Crops in Agriculture. *Advances in Agronomy*. 36: 97-143.

Dudley, J.W. 1993. Molecular Markers in Plant Improvement: Manipulation of Genes Affecting Quantitative Traits. *Crop Science*. 33: 660-668.

Eberhart, S. A. and Russel, W. L. 1966. Stability Parameters for Comparing Varieties. *Crop Sciences*. 6: 34-40.

Falconer, D. S. 1960. Introduction to Quantitative Genetics. Oliver and Boyd, Edinburgh.

Fasoulas, A. 1981. Principles and Methods of Plant Breeding. Pub.No.11. Aristotelian University of Thessalondki, Greece.

Fisher, R. A. 1918. The Correlation between relatives on the supposition of Mendelian inheritance. Trans. *Royal Soc. Edinburgh*, 52: 399-433.

Flor, H.H. 1971. Current Status of gene-for-gene concept. *Annual Reviews of Phytopathlogy*. 9: 275-296.

George Acquaah. 2007. Principles of Plant Genetics and Breeding. Black well pub., UK.

Griffing, B. 1956. Concepts of General and Specific Combining Ability in Relation to Diallel Crossing System. *Australian Journal of Biological Sciences*. 9: 463-493.

Gustafsson, A. 1968. Reproduction mode and crop improvement. *Theoretical and Applied Genetics*. 38:109-117.

Gustafsson, A., Hagberg, A., Persson, G. and Wiklund, K. 1971. Induced Mutations and Barley Improvement. *Theoretical and Applied Genetics*. 41: 239-248.

Harlon, J. R. And de wet, J. M. J. 1971. Toward a Rational Classification of Cultivated Plants. *Taxon*. 20:509-517.

Hayward, M. D., Bosemark, N.O. and Ramagosa, I. 1993. Plant Breeding, Principles and Prospects. Chapman and Hall, London.

Jenkins, J.N. 1982. Breeding for Insect Resistance. In K.J. Frey (ed.) Plant Breeding II. pp. 291-308. Kalyani Publishers, Ludhiana.

Jennings, P.R. and Herrara, P.M. 1968. Studies on Competition in Rice. II. Competition in Segregating Populations. *Evolution* 22: 332-336.

Kempthorne, O. 1957. An Introduction to Genetic Statistics. John Wiley & Sons. Inc., New York.

Khanna-Chopra R. and Sinha, S.K. 1998. Prospects of Success of Biotechnological Approaches for Improving Tolerance to Drought Stress in Crops. *Current Sciences*. 74: 25-34.

Lewis, C.F. and Christiansen, M. N. 1981. Breeding Plants for Stress Environments. In K.J. Frey (ed.) Plant Breeding II. (First Indian edition). Kalyani Publishers, Ludhiana.

Lewis, D. 1942. The Evolution of Sex in Flowering Plants. *Biological Reviews*. 17: 46-67.

Maheshwari, S.C. Maheshwari, N., Khurana, J.P and Sopory, S. K. 1998. Engineering Apomixes in Crops: A Challenge for Plant Molecular Biologists in the Next Century. *Current Sciences*. 75: 1141147.

Mangelsdorf, P. C. 1952. Evolution under domestication. *Amrican naturalist*. 86:65-77.

Mather, K. 1943. Polygenic Inheritance and Natural Selection. *Biol.Rev.* 18: 32-64.

Mather, K. and Jinks, J.L. 1977. Introduction to Biometrical Genetics. Chapman and Hall, London.

Mather, K. and Jinks, J.L. 1982. Biometrical Genetics, 3rd edn. Chapman & Hall, London.

Maxwell, F.G., Jenkins, J.N. and Parrot, W.L. 1972. Resistance of Plants to Insects. Adv. Agron. 24: 187.

Mehta, T. R. 1986. Early History of Plant Breeding in India. *Indian Journal of Genetics*. 46:1-11.

Michelmore, R. 1995. Molecular Approaches To Manipulation Of Disease Resistance Genes. *Ann. Rev. Phytopath*. 33: 393-427.

Murashige, T. 1974. Plant propagation through tissue culture. *Ann. Rev. Plant Physiol.* 25: 135-166.

Phundan Singh, 2006. Essentials of Plant Breeding . Kalyani Publishers, New Delhi.

Poehlman, J.M. and Borthakur, D. 1995. Breeding Asian Field Crops. Oxford and IB H Publishing Co., New Delhi.

Rana, R.S. 1995. Intellectual Property Rights: Protection of Plant Varieties. In R.S. Rana, P.N. Gupta, M. Rai, and S. Kochar (eds.) Genetic Resources of Vegetable Crops: Management, Conservation and Utilization, pp. 421-427. NBPGR, New Delhi.

Ranhawa, M.S. 1969. A History of Indian Council of Agricultural Research. ICAR, New Delhi.

Reddy, B.V.S. and Comstock, R.E. 1976. Simulation of the Backcross Breeding Method. I. Effects of heritability and Gene Number on Fixation of Desired Alleles. *Crop Sciences*. 16: 825-830.

Robinson, R.A. 1973. Horizontal Resistance. *Reviews of Plant Pathology*. 52: 483-501.

Robinson, R.A. 1973. Vertical Resistance. *Reviews of Plant Pathology*. 50: 233-239.

Sharma, J. R. 1994. Principles and Practice of Plant Breeding. Tata McGraw-Hill Pub. Co. Ltd., New Delhi.

Sidhu, G. 1975. Gene-for-gene Relationships in Plant Parasitic Systems. Science Progress, Oxford, 62: 467-485.

Simmonds, N. W. 1991. Bandwagons I have known. TAA Newsletter, Dec. 1991:7-10.

Simmonds, N.W. 1979. Principles of Crop Improvement. Longman, London, New York.

Singh, B.D. 2006. Plant Breeding: Principles and Methods. Kalyanai Publishers, New Delhi.

Singh, D. P. 1991 Breeding for Resistance to Diseases and Pests. Springer-Verlag, Berlin.

Singh, H.B. 1962. Exploitation of Hybrid Vigour in Vegetables. ICAR Series No.33.

Singh, R.K. and Chaudhary, B.D 1985. Biometrical Methods in Quantitative Genetics Analysis. Kalyani Publishers, New Delhi.

Sinha, S. K. and Khanna, R. 1975, Physiological, biochemical and genetic bases of heterosis. Adv. Agron. 27: 123-174.

Van Der Plank, J. E. 1968. Disease Resistance in Plants. Academic Press, New York and London.

Vavilov, N. I. 1951. The Origin, Evolution, Immunity and Breeding of Cultivated Plants. Chronica Botanica, Waltham, Mass, U.S.A.

WIPO/UPOV 1990. Report of the Committee of Experts on the Interface between Patent Protection and Plant Breeder's rights. WIPO/UPOV/CE/I/4, February 2, 1990.

Wright, S. 1921. Systems of Mating. *Genetics* 6: 111-178.

Index

D

E

F

G

H

I

L

M

T

U

V

W

www.ingramcontent.com/pod-product-compliance
Ingram Content Group UK Ltd.
Pitfield, Milton Keynes, MK11 3LW, UK
UKHW021451280726
14060UKWH00001BA/345

9 789387 057234